Further praise f

Barbara Hurd gives voice to the grief that anyone must feel who is alert to Earth's unraveling. She places our age of climate disruption in the grand context of geological history, with its record of five previous mass extinctions, and shows through accounts of her wide-ranging journeys what the cataclysm looks like up close, in calving glaciers, shrinking polar icecaps, dying corals, drying wetlands, diseased forests, plants unmoored from their pollinators, animals out of synch with their food sources, species millions of years in the making vanishing as we watch. Equally at home in the language of science and poetry, Barbara Hurd is an eloquent guide to the Anthropocene, the sixth mass extinction we're witnessing now, the one of our own making.
  – Scott Sanders, author of *The Way of Imagination*

Barbara Hurd knows the sorrow of our mortal condition and Earth's losses due to extinction. *The Epilogues* is a full-throated expression of grief framed by her keen attention to the smallest among us, the bizarre, grotesque, and beautiful--lungworm, carnivorous sundew, and fairy shrimp. No creature great or small escapes the fate of climate change. And yet Hurd gives us as well the big story of life on Earth, tracing all six great extinctions and the reinventions that followed. While sorrow here becomes "a dignified and valid emotion," loss is embedded in a sprawling web of mutualities always in a state of becoming a new story.
  – Alison Hawthorne Deming, author of *Zoologies: On Animals and the Human Spirit*

Also by Barbara Hurd

Nonfiction

Listening to the Savage: On River Notes and Half-Heard Melodies
Tidal Rhythms (with photographer Stephen Strom)
Walking the Wrack Line
Entering the Stone
Stirring the Mud

Poetry

Stepping into the Same River Twice (with artist Patricia Hilton)
The Singer's Temple
Objects in this Mirror

the epilogues

Afterwords on the Planet

Standing Stone Books is an imprint of **Standing Stone Studios,**
an organization dedicated to the promotion of the literary and visual arts.

Mailing address:
1897 State Route 91, Fabius, New York 13063

Web address:
standingstonebooks.net

Email:
standingstonebooks@gmail.com

Distributor:
Small Press Distribution
1341 Seventh Avenue
Berkeley, California 94710-1409
Spdbooks.org

ISBN: 9781637609132

Library of Congress Control Number: 2020951685

Book Design by Adam Rozum

Standing Stone Books is a member of the Community of Literary Magazines and Presses
Clmp.org

for Stephen

Barbara Hurd

the
epilogues

Afterwords
on the Planet

Allow, because you must allow, the broken glass to speak. "
– Carole Maso

He who seeks naively to embrace his own time will accept
its masks and illusions.
– Loren Eiseley

The greatest poverty is not to live
In a physical world, to feel that one's desire
Is too difficult to tell from despair.
– Wallace Stevens

Deep autumn ~
My neighbor
How does he live, I wonder
– Basho

Acknowledgments

Grateful acknowledgment to the editors of publications in which some of these essays, in one version or another, first appeared:

"Dead-End Hosts" was published in *The Ocean State Review* (Fall 2020).

"Epilogues" was first published in *Literary Hub* (June 2019).

Portions of "Uncertainties," "Sundew," and "Handholds" first appeared in *Invitations to the Indefinite* (Barbara Hurd and Meg Ojala, Flatten Art Museum, St. Olaf College, 2018).

Snippets of the Lamentations appeared first in *Terrain.org* (December 2017).

Portions of "Immersions," "Attachments," and "Exchanges" first appeared in *Tidal Rhythms* (Barbara Hurd and Stephen Strom, George F. Thompson Publishing, 2016) and were also excerpted in *Terrain.org* (October 2017).

"Glimpses" was first published in *Orion* (November-December 2016).

Heartfelt gratitude to the Guggenheim Foundation for a Fellowship that helped support this work.

And many, many thanks to the Thursday night writing group for 35 years of camaraderie; to Lawrence Raab, Judy Raab, Madeleine Deininger, Joel Peterson, and CJ Moll for love and support; to Margaret Gibson for careful reading of an early version; to Kenny Braitman and Ann Bristow for kinship; to my son Adam Wilson for keen scientific counsel, my twin Joanne Hurd and brother Ken Hurd for unwavering encouragement, my brother-in-law Chris Roe for traveling companionship, my daughter Tara Perry for laughter and joy;

to my grandchildren Caitlin, Samantha, Keva, and Asher for expanding my heart and for reminding me so poignantly why the future matters;

and to the love of my life Stephen Dunn for his discerning eye and generous heart.

Table of Contents

Introduction

Sometimes I think an earthquake might be easier to cope with. We'd be violently awakened, and the evidence would be instantly irrefutable. We'd know where to look, whom to call. It wouldn't be our fault. It wouldn't be anyone's fault. Even if we didn't know much about tectonic plates, we'd be pretty sure the catastrophe could not be blamed on human activity.

We'd know to tell the stories of both heroes and victims, whom to praise and whom to mourn, maybe even how to resume something resembling our pre-quake lives.

Instead, we are, in 2020, stuck in a morass of mounting collapses—seas inching inland, islands disappearing, more droughts, more flooding, more heat waves and storms, dying oceans, wiped-out species.

The climate crisis is complicated and sometimes subtle, without clear beginning or end, without a single cause we can point to and say, "There! That's the culprit." Some of us are often not even sure what—or even whether—we should mourn. The ongoing breakdown does not give rise to singular heroes or hope for redemptive endings. The ground is getting mushy or baked beneath our feet. Either way, our footing is more tentative now.

And when the foundation goes, so go the stories that supported them—and us.

If I call myself a naturalist and claim Thoreau was the one who got me started, I'd be guilty of clichéd thought and less than half correct. As a young child, I lived near Walden Pond, but the truth is my obsessions began as escapes from a suburbia that felt too small and therefore pushed me into the smaller still--nearby swamps of herons and frogs where the world grew large with life and death and the webby give-and-take between.

It would be decades before I read *Walden* and his *Journals*, years before I miscopied the line that cinched the essence of my obsessions, almost half a century before I understood a warming planet threatened not just individual creatures but whole species and entire ecosystems and therefore big swaths of multiple ways of life.

In one of his journals, Thoreau wrote, "Who placed us with eyes between a microscopic and a telescopic world?" But when I copied Thoreau's phrase into my notebook recently, I inadvertently left the "p" off the word "placed." "Who laced us here?" I wrote. Such mistakes have the power to subvert habits of thought and open new associations altogether

And that slip did, leading me to think not about "who," but about feeling "laced" between two worlds, torn between looking for big patterns and fingering the nitty-gritty, whiplashed between the scales of bullfrogs and the Sistine Chapel, close and far, intimate and beyond, now and next, all against the fathomless elsewhere of geologic time.

How to look at the world under a rock and from the top of mountain, simultaneously? Maybe someday I'll learn how to lay a large-scaled world on top of a small-scaled world without obliterating what's underneath. It's what restoration artists do with ancient palimpsests: make visible not just what's on top but what's begun to vanish underneath.

It's what storytellers do, too: lace the twined worlds of intimacy and perspective that are, finally, the homes in which we characters fashion our lives.

Exploring the effects of large-scale changes on small-scale creatures from the tropics to the Arctic has landed me, ultimately, in a deep uneasiness about the stories I grew up with. And out of that uneasiness an urgent question has intensified: If the big, instructive stories of western cultures (Genesis, Jesus as savior, Greek and Roman epics about singular heroes triumphing over their enemies) are losing the power to support us in times of crisis, then what will replace them? What kinds of new stories will help us navigate through the difficult times ahead?

The words in this book emerge from my restless urge to have one eye on the marvels of minutiae and the other on epic events and the power of story. There are fairy shrimp and lungworms in these essays; there are cataclysmic volcanic eruptions and asteroid collisions. And holding the tension between big and small, there is the human heart, groping for a new narrative, often unsure where to be and what to feel and how to say it.

Some of those uncertainties give rise to lamentation, what Wallace Stegner calls "the conscience speaking," which I might amend to say that lamentation is the conscience *struggling* to speak. One's conscience, if it speaks at all, more often wrestles to imagine a voice not borrowed but invented, perhaps shaped into rhythm and song or into hair-pulling, clothes-rending, god-appealing wailing, as the women of Ur did. Some 4000 years ago, their city in ruins, they wandered speechless, as if their tongues, too, had been stilled. And then they invented their own language for grief.

For many of us who live in western cultures, raw displays of grief are often looked at askance. They're considered unrestrained outpourings, uncontrolled keening. We who are steeped in mannerly ways and rituals grieve with decorum. We sit shiva, stage solemn funerals, orchestrate rites of passage, fill chapels with flowers and hymns and eloquent eulogies. Well-schooled in ceremonial displays of grief, we're counseled to "work through" our sorrows and save our real grief for behind closed doors, where it can subside or transform or fester and consume.

But the devastation we—every creature on this planet—face today is no longer local or personal. Soon, it will be undeniable. It will break through closed doors and private heart-work. It will shatter secret dreams and community traditions. Though some will be disproportionately affected more and sooner than others, the climate crisis will, already has, affect us all, and if our most respected scientists are right, we have, in many ways, already doomed our planet.

If personal grieving deepens and spreads into collective, world-wide grieving, then what forms will our aching take? What sounds our bereavement?

Plato condemned lamentations. Let the disreputable wail, he insisted, not the decent, the noble. He censored lamenters for the same reason he banned poets: Both can express those shadowy feelings that might distract us from pursuing the idealized forms he saw as ultimate truths. Both can praise the richness of transience and can prohibit platitudes, purities, tidy solutions that sabotage the need to ponder the impetuous in our hearts.

Neither one will halt what's happening now.

When Robert Frost was asked whether art, to be considered good, must prompt some kind of action, he must have paused before he asked, "How soon?"

He's right, of course, that art does not a revolution make, especially in the little time we have left before we reach some critical tipping point. And however soon "How soon" might be is likely way too late for us, except that we might, in the meantime, shift the ways we grieve what's going and what's gone. And since those losses are both personal and global, then maybe the urge to speak, to shape, to share whatever lamentations begin to emerge can change how we respond to one another and to this home we're now destroying.

I hear the collective voice in the laments that follow as a chorus. I come from privilege; many who carry on the traditions of lamentation do not. But the voices of grief, though they take many forms, erupt from a place that is, I believe, shared by us all, one that does not belong to any one culture or race, country, even era, but to the rich and wretched truths of our humanity—or lack thereof. Because they arise out of my own semi-conscious, they are inevitably tinged by the culture of my upbringing. Dare I say, though, that they seem to break out from some fractious undercurrent not just of my own background but of some common heart-thread that binds us all? They are raw voices, often communal, always marginalized. Cantankerously impatient, sick of explanations and excuses.

I recognize them and I don't.

In the pages that follow, their chorus begins to infiltrate current stories of decline, especially those that threaten creatures that need water or mud, meaning all of us.

Hovering in the background, the five mass extinctions recorded in the fossil records of deep time remind us that loss is universal and timeless.

So are our bungling attempts to cope with it.

Preamble

The beginning is so far back it can only be known as before-the-beginning, meaning it began in mystery and nothingness and continued that way for billions of years while there was only gas which took a long time cooling and coagulating until eventually a surface skim became crust.

Thick clouds sheathed the hot earth for long, long time. Any occasional cloud-drip down toward the still-molten planet hissed right away and steamed back up. Picture water sprinkled on an ancient white-hot fire. It took another few billion years for the planet to cool enough for drips to begin to drop as rain and once that rain began, the clouds dumped it down day and night for several hundred million years, filling basins with warm shallow waters in which nothing lived.

To speak of that lifeless time requires words like prior and before, onset, origin, prelude, preamble.

Add plotless, static, stalled, all of which mean the story had barely begun.

There could be no plot because other than a long, slow cooling and then torrential rains, there was no action, no sign of any future that was any different from what already was. And so it was for 90% of the planet's history.

Paleontologists call this before-the-beginning time the "boring billions," endless eons of rain, rain, rain.

And then a hiss from something we might call alive. Maybe it was some inexplicable shrug or churn of methane, maybe CO_2 and a chemical slosh one way and not the other, but the result was a haphazard mix that at some point became a single cell, neither plant nor animal, oozing somewhere between alive and not, and then another and then blue-green algae appeared and then a billion years ago—worms, jellyfish, sponges and things that slurped up slime and seafloor goop. And on and on it went: three-lobed trilobites, mollusks, small cephalopods whose remains we know only from ancient fossil records.

And then, some 500 million years ago, most of them vanished: 75% of those trilobite families disappeared. 50% of the sponges, most of brachiopods and gastropods—gone, wiped out for reasons we don't know.

For many of those primitive creatures, that was their end. For them, the first round of experiments had failed.

Underwater

The sea subverts all stories, including its own. Trying to describe a recent trip to Hawaii, I fumble for highlights, attempt to reconstruct a few adventures--one granddaughter had learned to snorkel, the other to dive through waves. We were privileged to have made the trip, but re-telling little triumphs does nothing to evoke both the magic—feeling delightfully lost in warm waters--and the tragic--feeling unnerved by the sight of bleached coral and eroded beaches. Every conscious attempt to storify the experiences peters out. Why is it so hard to convey the tangled emotions, the worries, even the thrills?

Maybe the problem is partly a matter of scale: all that water, those scattered specks of land, we teeny human beings. To swim there among remnants of volcanoes is to be reminded of billions of years preceding this small slice of the present, of the previous extinctions and climate catastrophes, along with the planet's ability to regenerate, start anew, those antidotes, perhaps, to the crisis mentality that consumes so many of us now alarmed by climate change. The long view, after all, usually puts problems in a different perspective. Against that ancient history and the inevitability of eternal flux, trying to tell a riveting story about our mere week anywhere has seemed a little pointless.

But drifting those days among coral reefs and yellow tang, I became aware of a growing need for a fiercer, more focused kind of response. Maybe it's because the climate crisis means real trouble now in those waters, along the shores, in the valleys, trouble on the ice floes, in the reservoirs; people are starving, species disappearing. Maybe it's because the casualties today seem freakish, unnatural, preventable. Among kelp forests and coral and drowning tidal pools, dead lakes, warming bogs, and melting glaciers, *remember the big picture* seems like an increasingly lame consolation.

And maybe it's because being in warm waters invites a different—more fluid?--sense of what matters and why and how to say it. To address the climate crisis, maybe what we need now isn't just more committed politicians, more science, more accountability, but different kinds of stories.

What if we began with Akiko Busch's premise that ". . .the measure of our humanity may be derived not from how we stand out in the world, but from the grace and concord with which we find our place in it." Not from our own heroic epics but from how we embed ourselves, fit into larger patterns. Like so many tales from indigenous peoples, such a story would mean refusing to be moved mindlessly along or above, trying instead to stay stilled and saturated in a world that was long ago ours and not-ours. It would mean learning to see again what and whom we live among, to immerse ourselves *here* so deeply our chiseled sense of "I" begins to soften within the vastness of the universe.

The poet Muriel Rukeyser says, "The universe is made of stories, not atoms." But atoms can be stories also—too small, perhaps, to become legends, too fast to serve as parables--but stories, nevertheless, of attraction and repulsion, instability and cohesion and the power of invisible elements to combine and shape the world. It's those less visible elements I've become interested in now, the ones that often don't make it into epic stories. Among the well-shaped tales, the small, oddly shaped moments—more radial perhaps than linear--fade. And that can mean they're gone. Essentially erased. And yet often such moments remain under the surface, invisibly shaping our lives as the barely detectable perceptions juxtaposed against a wisp of feeling. Impressions that crystallize in the gaps between legends, like those strange moments in Hawaii when we snorkeled one afternoon and, with our heads underwater, the horizon vanished and there was just blue above and blue below, and my granddaughters drifted closer, wanting to hold hands as we dipped towards the sea floor alive with fantastic growths all spiny and twisted while through it all swam hundreds of striped and speckled fish.

You get pulled in. You fall in love. If you could weep behind a snorkel mask you would. And you want to tell them that if we were strong enough swimmers we could swim southeast and then east, then north through each of the seven seas which are of course not divided from one another, there being no walls underwater, no customs officials, no gates, no change in language or currency or clothing, none of those ways of putting unruly things in order. And then one of them squeezes your hand and the other nods and you somersault in slow motion, all three of you, and something about the rhythm and the holding hands and the turning upside down puts you on the verge of returning—to what?—spaciousness, fluidity, a chance of going too far, all the way around, only you don't mean around the globe, you mean around constraints, the useful boundaries of good sense and for a few minutes you want to stay there, forever, with the two of them, in a world far larger than the one you usually live in. This is a different kind of story. This is a response. A painting in motion. It was a small moment, maybe two or three minutes, but it is what has stayed with me ever since.

I remember little else about the swim that afternoon in Hawaii—not the name of the cove, the rest of the day, not the way those moments might have marked a pivotal point in my life. A therapist might want me to probe it, figure out how it played into the development of my psyche. A friend might want me to turn it into a story about what happened next. I'm not interested. I want only to hang on to the visceral memory of being together in that dark ache and blue water and to acknowledge the truth of our lives as a series of moments that often don't add up to epics but that might, nevertheless, reverberate through the years. And to be reminded that it's resonance like that that helps signal *home*. And lack of resonance that can alert us when our lives have become atonal, shrill, shriek-like, or just disturbingly out of tune.

A few days later, SNUBA-diving off the Kona coast, I asked my guide to show me more corals. My tank was fastened to a small raft that floated above me, up on the surface of the water. I was connected to that vital oxygen by a tube that trailed after me, like a long umbilical cord. The deeper I went, the harder it got. The pressure underwater was both literal--in my ears and in every cell of my body—and metaphoric—in the urge to elongate and glide, to prolong the sensations of absorbing and being absorbed. I dropped ten feet, fifteen. The water got murkier. The world above grew farther and farther away. And then the guide appeared in front of me and gestured ahead where clump after clump of *Pocillopora meandrina*—cauliflower coral—squatted on the ocean floor in a pointillist swath of pale greens, lilac, pink. In each one, branches radiated from the center, the origin of its growth. The formation is named for its resemblance to a head of cauliflower and when I swam closer what I saw did indeed look like cauliflowers with their clusters of florets flattened and pried slightly apart. Small green fish darted in and out the crevices, then a couple of sleek black ones, and then three turquoise, gorgeous against vibrant pink and through all of that the question of whether seeing all of this can ever be enough. What if this intense looking is just more of that try-to-be-attentive marveling I've done for so many years? What if I've let wonder and awe excuse passivity?

* * *

Swimming over the reef, my guide suddenly twisted toward me, extended his arm palm up, wiggled his fingers fast, and pointed. It's a standard warning sign: Danger: venomous sea urchin. Don't go near it.

That's the kind of hook, I realize, despite my recent leanings, which could begin an important story.

Drifting a little closer, I saw the sea urchin had tucked itself into a coral reef crevice where it resembled a pincushion bloom of blue-black, glassy spikes. *Echinothrix diademae* lives in waters up to thirty feet deep all over the world. It moves by hydraulic pressure, squeezing water in and out of its many tube feet, which are not visible under the profusion of spines that covers the body. Its mouth is not visible either. It is, in fact, positioned on the creature's belly, while its anus rides on top. Consisting of five tooth-like structures that Aristotle described as lanterns, it's a mouth that some hope might one day help to save the coral.

Inside the skeletal coral lives the animal itself--a translucent polyp--and inside the polyp live millions of many-hued, single-celled algae—zooxanthellae—which spend their whole lives sequestered there, dazzling the reef with color and doing what all algae do: producing oxygen, glucose, and amino acids, which the polyp needs to stay alive. In return, the polyp provides protection and the compounds the algae need for photosynthesis. Coral is one of our oldest animals—500 million years old. Algae is even older. Their symbiosis has been stable, mutually beneficial, for over two hundred thousand years.

* * *

The current climate crisis is altering the reef. Some unknown trigger in warmer, more acidic waters can cause the polyp to expel the heat-altered algae it depends on. If this were a human story, would we call the main character self-destructive or self-protective? If what we ingest is damaged, should we expel it, even if it's what keeps us alive? But the polyp has no brain and can do nothing to halt the process. The crucial

26

symbiotic relationship is disrupted. Without the internal algae, the polyp is colorless; the coral itself grows pale—bleached--and weak. Coral-building is crippled, and the structures that do manage to survive grow less impervious, more susceptible to breakage, an underwater array of antler-shapes and fans, fingers and boulders, all of it osteoporotic and frail. If the reef finally dies, with it go the sponges and clownfish, sea stars and crabs, the thousands of marine creatures that count on the reef for protection and food.

And the scenario gets even worse: warmer waters and a less robust reef can trigger the growth of macro algae, which, in moderation, can benefit the reef, providing food for nibbling fish and helping to cement the structures. But if nutrient levels are high, populations of big algae explode and smother the coral, blocking sunlight. Its holdfasts can clog crevices, its filaments can strangle delicate life forms. Swaying in the currents, large algae—ropy strands and feathery plumes—brush over the polyps, causing the sensitive creatures to quickly retract into the safety of their calyx cups. Too many sea stars, too much retraction, not enough sunlight, algae-borne pathogens: the effects cascade and the reef begin to collapse. Once described as the "rainforests of the oceans," home to a quarter of all marine species, the acres of once-vibrant coral turn bone-white and bare.

With its five teeth, perpetual hunger, and the aimless grazing that is its life, urchins love big algae. One species—*Diadema antillarum*—is particularly prodigious. It cruises the reefs, using its teeth to scrape up the macro algae. Researchers even raise urchin larvae in labs, transplant them to an "urchin-enhanced reef" and watch to see if they'll take up the job of scraping the corals clean, letting their insides show through again. Hope, as they say, springs eternal.

So, I have often thought, does the value of venerable narratives. But having drifted above that magnificent reef, relishing the sight of the not-yet-ruined, I worry that epic tales and their instructive highlights of how the hero did or didn't overcome the odds can too often be tales from which the truths of other lives have been omitted. And those omissions, increasingly disastrous, can keep us—especially the privileged us--out of the untold troughs and rougher waves of heart-twisting grief.

If I were thinking only of the stories I grew up on, I'd say beware of such troughs; they can create the stasis no plot can afford. But I'm not, and below me that day the spines of that urchin were long and needle-sharp. In other places of the world, this species is going bald. Their spines are dropping off, which leaves more of the urchins defenseless. Death rates are up. In the Cook Islands, thousands of dead ones have washed ashore. Scientists suspect a pathogen multiplying fast in warmer waters, perhaps something similar to the one that killed over 90% of *Diadema antillarum* in the Caribbean a few decades ago.

As is true here in parts of Hawaii, the coral reefs around the Cook Islands are in trouble. They're in trouble in Australia, in the oceans off East Africa, Florida. Almost half the world's reefs are in decline. By 2030, scientists estimate, that number will approach 70%. Two decades after that: 90%.

In front of me, my guide spotted a school of triggerfish darting ahead and excitedly motioned me on. Instead, I descended a little more. What if I removed my flippers and kept dropping until I could feel the glassy spikes of that blue-black urchin press into the flesh of my bare foot? Maybe that's what I mean by a fiercer, more focused kind of response. Not the sometimes crucial, sometimes mindless, knee-jerk reaction of forever trying to rise up and move on, to make progress, but the pain, every day, of dropping down and into. The visceral knowing that comes from being *among*. Bigger, older stories might show us a moral way forward, but we need the smaller ones, too, the ones that emerge from lowering ourselves back into this physical world where a lament might, as Carole Maso writes, "allow the broken glass to speak." What if, in other words, we let destruction show its inarguable way.

Lamentation

1

*Because the part of me with a breaking heart can fixate on final moments, I
keep reciting a litany of lasts—last breath, last song, last supper,*

*last pinta tortoise's closing hour on this planet, last thing the Caribbean seal
monk saw as it surfaced one final time off the coast of St. Croix,*

*and then imagining last chances: the strained grunts of some about-to-be-
extinct creature's final mating—do any of us imagine that?--last looks, final
touches, last hope for keeping the species alive?*

*and more of last suppers: turtle eggs, insects, grasses of the Galapagos, last
gestures, last sounds, last moments on earth, last folding of the wings.*

*Breaking hearts are not in the business of writing epitaphs. They want no
summary, no story, no final honoring of the dead. They want the great auk
squawking its own requiem, the final coos and clucks of passenger pigeons
too weak to fly, perhaps to hear for the first time the last breath of the
last dodo bird*

*because how else but by breaking again and again can a surviving heart
learn that when it comes to death, practice doesn't count or alter the end,
does nothing but affix the heart to a point just shy of the finish line where
the claustrophobia of life up-close next to death up-close can leave us
immobilized or maybe roused at last to try to reverse a final countdown.*

Watery Beacons

Swimming thousands of miles south of the Hawaiian reefs, I was not thinking about story or death; I was, instead, enjoying the relief of being in the water. A few minutes earlier I'd cringed my way across the volcanic shore, rock-lava hard and sharp on bare feet, crossed a narrow strip of sand, and waded into equatorial waters off the Galapagos Islands. The comfort was instant: my feet stopped stinging; I felt immersed, lissome, my body buoyant, every change in direction sinuous and supple. A green sea turtle, unaware of either story or sinuosity, swam just a few yards beside me, nibbling algae off the coral. Both of us banked and dove, glided, rolled one way or the other, moved on a three-dimensional, underwater stage like ballerinas, she with a heart-shaped shell on her back and I with a question about whether one can ever move gracefully in inhospitable spaces. Is it possible to be at home in a place that doesn't feel any longer like home?

A visitor in this realm, I am a good swimmer but, after millions of years of evolution, made for walking on dry land. My kind knows how to make almost any habitat hospitable. We know how to shelter and feed ourselves as we move from shore to forest to desert to tundra. We even manage, now and then, to do so with a modicum of ease. We keep ourselves upright and avoid collisions, even swing our arms as we stroll, though they're not necessary for locomotion.

A turtle on land is neither graceful nor safe. Having bided her time just off-shore, a gravid female waits until the safety of darkness to lumber ashore. She swishes her flippers, trying to push off dry sand that often just gives way beneath her big-domed, two-hundred-pound body.

If she manages to drag herself up the sandy slope, she'll spend two to three hours digging a nest, using her back flippers to scoop and fling sand until the hole is at least a foot deep and she can drop a hundred eggs.

In the darkness, crabs might be watching, maybe a shorebird or two. But safety—her own--evidently trumps maternal instincts, for as soon as she finishes patting sand over the nest, she abandons it. Turtles in a hurry are almost comic. Unburdened now by eggs and aided by the downward slope, she flippers in the sand as fast as she can down to the surf and disappears into the safety of water where she resumes her gliding and nibbling. The turtle swimming near me that day might, for all I know, have just, hours earlier, risked her life to start some new ones and was swimming then with what passes for turtle relief.

And back up in the sand, the eggs, if all went well, lay undisturbed for two months.

This is the scenario that's been playing out for a hundred million years. She lumbers ashore, deposits eggs, lumbers back, re-enters what is, for her, the more hospitable home. The eggs hatch, the hatchlings waddle en masse toward the water.

Nothing, though, in the ancient history and re-playing of this on-shore interlude has prepared the sea turtle for changes brought on by rising seas and more violent storms. Recent hurricanes decimated Florida beaches, many of which had already been critically eroded. On a shrinking shoreline, nesting sites above the high-tide line are at a premium. How far up-beach can a turtle go before she's digging in the shadow of tiki decks and cabanas?

And even if she manages to nest in a safe spot, there's the problem of the temperature of the sand in which eggs lie incubating. Inhospitable, in this case, does not mean too cold. It means too warm. Turtle sex is not determined by DNA. When the female deposits those eggs, they're sexless. It's the surrounding sand that determines, over the next few weeks, which embryos develop as males, which as females. Sand temperatures below 83 degrees means males, between 83 and 85 degrees means it could go either way. Above 85 degrees means females. A few degrees hotter and the eggs will cook in their own shells.

It's not hard to predict the catastrophe that's pending. From the beaches of Australia to Florida across the South Pacific and Indian Ocean, the sand has already begun to warm, and the ratio of male to female has already begun to change. Recent studies off the Great Barrier

Reef reveal that hatchlings scurrying seaward from their sand-warmed nests are now 99% female. And though it's true that most populations need more females than males to sustain themselves, 99% is a whopping big margin. The species, some scientists worry, may eventually die off.

What kind of story can help save a whole species? Not, I suspect, the one based on the hero's journey. Heroism, Virginia Woolf once claimed, is a type of botulism, poisoning a society from within with stories of singular conquest so blindingly triumphant that the lead characters seem like super beings from another planet, nothing like the rest of us blokes, which means the rest of us blokes can shrug our shoulders, resign ourselves to the miasma of passivity.

But what if we had a different sense of hero? Not the singular figure rising above or collapsing below an adversarial nature, but the collective, multiple figures, a sense of inclusivity. Perhaps the community as hero.

It's not that I don't love a well-told story or the heroes who populate them. It's that I'm increasingly aware of how the grand stories so many of us in western cultures were raised on—Genesis, Moses parting the water, Christ rising from the dead, even Paul Bunyan's feats and Ahab's obsession—continue to spotlight heroic characters performing heroic deeds, overcoming—or not—incredible odds. How could they not dominate so much of our consciousness? They're grandly rich stories, instructive and soulful. It's just that their quests seem so dominated by large figures and so oblivious to the rest of us and how we actually live our lives. And if those heroic quests take place on post-Edenic stage that sets us up to long for a life beyond this blighted world, then they have little power, anyway, to help us re-imagine ourselves as beings enmeshed on this planet we're now destroying. And nothing now seems more urgent than that.

For turtles, shrinking beaches and skewed sex ratios aren't the only urgencies. As those hatchlings pip their way out of their leathery eggs, instinct tells them to head immediately towards the light. For most of the millions of years on this planet, the sea--silvery in moonlight, white crested with waves, palely lit by far-off horizons—has been that reliable beacon. The hatchlings wait in the nest for darkness and then head straight for it.

Imagine the hatchling's disorientation when, having pipped its way into the air and sounds and wide-open space of the world beyond its shell, the first lights it sees now come from nearby condos or boardwalk restaurants, cars on a pier, streetlights. The initial moments out of the shell are a turtle's most precarious. It can't see well until it's underwater. Defenseless on the beach, its only hope of surviving lies in making a fast scurry to safety. But if the watery beacon is obscured by artificial lights, the hatchling lurches awkwardly in one direction, then another. It can't get oriented toward the ocean, which is its only future. Befuddled, it may wander too long the wrong way, while dehydration, exhaustion, the swift swoop of talons or fast pass of a dune buggy take their toll. One percent survival rates, plummeting numbers of males, vanishing beaches—none of this bodes well for the species whose birth requires they lumber ashore.

The poet Michael Longley says that good art tunes us up, makes us smarter and more sensitive. Our hidden, buried lives momentarily align with someone else's so that we feel seen, even welcomed. Maybe hospitable places do likewise, only instead of tuning us up, they tune us in, help us feel in accord—but with what? Not just our obvious needs and a land's ability to provide, but with something deeper, less seen: an alignment with the subtle rhythms of a place, its pattern of yielding and resisting, to which we respond by accepting and ceding, the ancient give-and-take, critter and place coordinated, movement itself, in fact, not cumbersome but agile.

Turtles don't have art as we think of it. And if they're telling their own stories of survival, we don't know how to hear them. They have years of easy swimming in mostly hospitable seas interrupted by graceless treks to lay their eggs on beaches that are heating up and beginning to disappear. They survived the extinction that wiped out the dinosaurs. They might survive what's happening now. In the meantime, as those nesting sites grow more compromised and the hatchlings more disoriented, can they tell that their birthplace is becoming less hospitable, less like any kind of home?

Lamentation

2

All night, they keep coming back, urchins and coral, sea turtles and
yellow tang, hundreds of them dark blue, bright green, bright yellow,
silent and therefore ominous as they drift through a muteness I know I've
helped make and cannot undo. What would I want them to say? That
they're doing well and thanks for asking? Small-brained and wingless,
at the mercy of currents and the history of my kind's undoing, they keep
what passes for their faces pressed against what passes for mine as they
move through my dreams and the reefs and up through the warming
water to beaches strewn with more than castles collapsing in the sand.

Don't tell me again that death's just a pause before ascension, a story
among many offered mostly to console. Whatever ascends here at the
water's edge does so with stench and bloated belly.

I'll say instead that they keep coming back, keep turning in the murky
water as if to look straight at me, and that though I'm aware how
precious this sounds I wake up knowing that whatever story about
extinction remains to be told must be told now eye-to-eye, graceless and
blunt, must be true enough to rouse, must not tempt any of us with solace.

Glimpses

The head of a Galapagos tortoise seems like an afterthought, a slightly angled bulge with eyes and mouth at the end of a long neck. It emerges from its shell like a leathery periscope angling awkwardly for a better view of the likes of me, who was trying to pass it one afternoon on a narrow path on an undeveloped island in the middle of the Pacific Ocean. The neck and head skin is scaly and rough, its feet like an elephant's. But it was the shell that intrigued me at first. It was all I could do, in fact, not to reach out and rap my knuckles on the ridge. It's the same urge you feel on a beach when you pick up a clam or horseshoe crab. Knock, knock, you think. Who's in there? Among some of us in certain moods, everything can be a shell, meaning everything suggests something hidden and therefore poses the pestering question of whether to probe or to leave it alone. Among some of us in certain moods, the latter is not an option.

Dissection is not the next step. I've fingered cats' kidneys in biology labs, pried open a scallop to study its abductor muscle. I know there are ways of getting physically inside almost every kind of living thing and I know that inspection—taken etymologically--can brutally invade even as it satisfies a simple urge to see. I'm more interested these days in the imaginary invasions that begin with a glimpse and end with a wholly envisioned, but still hidden world.

Take the work of putting oneself—literally or imaginatively—companionably among other animals, in the Galapagos, for instance, where I almost stepped on a two-foot iguana perched, dragon-like, on some rocks I'd been climbing. It eyed me but didn't move. I sat down beside it; both of us looked out at the sea.

Before that, I'd been turning somersaults in the blue waters just offshore when a sea lion approached. I came back up out of my eyes-closed-underwater-spin to the surface, shaking the water from my hair and blinking in bright sun, and there it was, lolling right in front of me. I could have touched its head. The isolation of the islands means its creatures have never learned to fear us, which means, too, that we have to adjust our sense of self while we're there and feel how it is not to be feared, to be just one creature among many roaming around the ancient rocks. It was, I believe, a glimpse of a moment of what might once have been our past.

In the Galapagos, the iguana and the sea lion know nothing of our history of exploiting them for exotic pets, for fur, for the aphrodisiacal effects of their reproductive organs. Without even an instinct to flee, they act as if the recent past has not yet arrived and that we are not who we've become. There, in fact, I was merely a furless, two-legged creature, and so long as I didn't swoop down with talons and beak, I was as welcome as anything else to sit close by on a rock and study the sea.

Imagine that.

Earlier, I sat in a harbor-side café on the one touristy island, browsed the Internet, then boarded a bus, a ferry, another bus and then a boat which took me way west for hours to a lava shore. Disembarking, I waded through clear water to the island, headed inland, and came across the giant tortoise, likely a century old, lumbering along the same path I was, thirsty, I imagine, as I was, under the blazing equatorial sun. Headed for mud holes or for cacti to which a few drops of dew might be hanging, it's been hauling its reptilian body over the burned-out ground decade after decade.

Its ungainliness is almost embarrassing. Really? This is the best evolution can do? And yet there is something admirable, too, in its slow pace and patience. Down in the Santa Cruz cafes, I'd heard that dry-island tortoises will lick boulders for any overnight moisture. In some places, generations have been licking the same spot for so many years they've dimpled the stone, created a pock in which dew can pool. Mindless repetition, some might say. Perhaps, but also a glimpse, I might add, of what can be worn away if survival is at stake.

Consider, for example, the short-sighted view of time many of us humans have. Literal interpretations of the Scriptures tell us the earth was created in seven days a mere six thousand years ago and, except for the Flood, hasn't changed a wit since. According to recent polls, something like half of all Americans still believe that. People evidently had history, but the earth did not. Never mind those fossilized sea shells found on mountaintops. "Spontaneous generation," early naturalists explained. They just bloom, shell-shaped, out of the rock. The result of astral forces, others maintained. Or "wet exhalations," "lapidifying juices," or the result of salt exploding inside rocks, anything that didn't require imagining the earth's history as far more ancient than humans' or the Bible's. It wasn't until the 17th and 18th centuries that some began to see those shells as early glimpses of what we now call "deep time." The planet's ancient past, which for centuries remained so hidden nobody even knew it existed, is still being revealed, which means "creationism" is being eroded, little by little, one tongue on the rock at a time. Without the longer view, what's at stake is not just our daily bread but our planet's survival.

There on the Galapagos Islands, though, those glimpses are not glimpses and do not need imaginative leaps. The evidence is not hidden. If you're lucky enough to get there, it's everywhere in your face and under your feet: newly formed volcanoes, cactus-studded lava fields, barren islands with steam-leaking fumaroles, and mists that can turn a lava slope green with vegetation. The place feels primitive. And by that I mean you don't have to use clues to extrapolate backwards in time to what might have been here eons ago. Being there is like being in a filmstrip run in reverse. Flying away from the mainland, you see nothing for two hours but gray light and gray water everywhere. And then finally a speck of blackness and then another, thirteen major islands in all strung out across 150 miles in the middle of nowhere. Never a part of any larger land mass, they spouted from a geologic hotspot that shot lava-plumes through the overlying tectonic plate and into sea and light where they cooled, formed the islands, one by one. The plate, still drifting after millions of years, acts like a conveyor belt, carrying them east, where slowly, over millennia, they sink out of sight. Darwin first labeled the islands "infernal regions." Melville bemoaned their desolation. Their discoverer, a Spanish bishop named Fray Tomas de Berlanga, described them as a place "where God had showered the earth in stones. Worthless."

The Galapagos are an anomaly, volcanic islands largely undisturbed. The oldest ones were formed some three to five million years ago, the youngest just a million ago. Newborns, I thought, as my plane descended and flight attendants sprayed the cabin interior with pesticide in case any of us carried miniscule contaminants in our clothes or hair. They'd already searched our luggage, queried us about carrying fruit or plants. Nothing, they're determined, from the outside microbial, plant, or animal worlds—except us-- would be allowed to enter this protected place. At the Galapagos airport, I walked across a welcome mat soaked with pesticides.

The imagination, however, cannot be kept out. Before I'd met the tortoise on the path, I'd been standing by a small salt-water lagoon drained and fed by slow slosh from the sea. The scene suggested something, not just about the connected bodies of water, but about ourselves and our ceaseless slipping edges that are, at our most fundamental level, who we are. When I squinted my eyes, the pintail ducks and blue-footed boobies could have been islands far out at sea or planets in a star-strewn galaxy. I myself might have been floating among them, all of it and all of us about to dissolve as colors shaded and light shifted in the shimmery heat. In many ways, the lagoon is the opposite of the Galapagos, whose volcanic landscape is harsh. But the lagoon and lava and sea all around can help us know the pleasures and perils of feeling immersed in the boundless— in place *and* in time—and thus permit us to get a glimpse of ourselves as we might not otherwise see us.

Or to get a glimpse of the world without us at all. In the Galapagos we can imagine landscapes that remained humanless for millions of years while endemic species, without competition, sprouted newer species and on and on the process has gone through thousands of years of isolation on islands exiled from the nearest continent. With plenty of food and almost no predators, what's to stop a thing from growing? Here tortoises grow to the size of small hippos, beaks change shape, wings atrophy, iguanas feed in the sea, the myriad creatures mutating and veering into unoccupied ecological niches uncrowded by other species in a purgatorial landscape of fumaroles hundreds of miles out in the sea until Darwin had enough evidence to formulate the whole shebangy theory of evolution.

It was a scientific vision in his case, not an artistic one that grew from initial glimpses and ended up repositioning us, once we finally emerged on the planet, away from the center of the universe, left us marooned with no God-given blessings or innate privilege, left us with the responsibility of intelligently designing our own moral lives. Boundless, too, that perspective. And disconcerting.

* * *

Of course things evolve everywhere and all the time. But sometimes the rare and new can only emerge—or be recognized--in isolation. Beginnings can be tender, in need of undisturbed space, lack of predation, nothing to flee from. We can think of those places, as Darwin did, as laboratories where lack of interference means a recent mutation or fledgling idea has a chance to grow. Or as Montaigne does, putting the image in human terms: "We must reserve a little back-shop, all our own, entirely freed, wherein to establish our true liberty and principal retreat and solitude."

Being "en-isled," I used to call it. Back when I was young and overwhelmed by what I saw as too many intrusions on my solitude, I had a whole theory about being en-isled. It pertained to only parts of us-- which parts I could never say--but the idea was that some parts of us get siphoned off somehow and remain sequestered on a remote island somewhere inaccessible, except to the imagination, which would, for no apparent rhyme or reason, go there now and then. There were no direct flights to this island, no way to will yourself there or back. You just woke up some mornings and there you were. If it happened to be on a busy day full of other people, the remoteness might scare or relieve you. If it happened on a quiet evening, it wasn't absence you felt but presence, and sometimes the fruit of imaginary palms was all you needed.

I don't mean to suggest I'd ever want to forgo the richness of human company or accomplishments in order to decamp to some island. Live entirely without others or without ever having seen great art or read great books? How else could I learn to think about the allure of the ill-defined and boundless? How else could I know that en-isled part of myself or grow alert to its satisfying dangers? And though I knew even as a child that contact with the bigger world is how we fine-tune our sense of self and fashion decent ways to live, I knew, too, that even an occasional glimpse of such an interior island could encourage new sproutings and experimental veerings—some healthy, some likely not. As with whole new species, with plenty of food and no predators, what's to stop a thing from growing?   There would be time later to haul such things off the interior island, let the actual world test them, as it will.

On the Galapagos there's been little off-island testing. Or, if there has been, we're unlikely to know much about it, as such testing would have meant extinction, and likely pretty fast. The Galapagos cormorant, for example, wouldn't last long in any other environment. Largest of all cormorants in the world, it has, like all of them everywhere, webbed feet, a hunger for fish, and a habit of spreading its wings after every dive in order to dry them. But for the last ten thousand years what the Galapagos cormorant spreads is shrunken, vestigial, more like feathered stubs. Why waste energy growing wings on an island where abundant food and lack of predators means what matters is sea, not sky? Who cares that you were a bird who was supposed to fly and now no longer can? You grow a strong lower body for kicking underwater, and don't worry about definitions. Earlier in the day, I'd stood on ankle-twisting, hardened lava beds and watched a cormorant waddle to the rocky edge and plunge. Going wingless was a gamble, a throw of the dice on the supposition that the sea would provide and that predatory eagles and raccoons would never make it to such an outpost. So far, it's a gamble that's paid off. Still, is it luck? What to call those accidents, those aberrant chances-become-changes that don't emerge from clear vision?

Scientists would call them evolution. Artists might call them opportunities. Wise Ecuadorian officials called them an obligation to impose a vision on what had already happened. Perhaps they knew German philosopher Friederich Schlegel's claim that the test for a completed work of art is whether it is "limited at every point, and yet within its border is without limitation." Perhaps the officials saw the archipelago as a work of art they had a chance to protect. In 1998, they passed protective laws and began to restrict visitors. Later, they imposed maximum weights on baggage, tabooed certain items, and began spraying the interiors of airplanes. There's talk even now of tightening those regulations. Perhaps that's the paradox of both the archipelago and my childhood sense of being en-isled: both resonate beyond boundaries and neither would be possible to sustain without limits.

* * *

The work of using a glimpse to imagine the hidden whole is the work, too, of story and language, of naming a barely felt feeling and watching it expand through phrase and rhythm or nuance and shape into something akin to the experience itself with all its undertones and complexities. "To glimpse" originally meant "to shine faintly." It's part of an artist's job to buff that faint shine a bit, bring it closer to visibility. It's part of our human job to wonder what's been obscuring it in the first place—what fears or faiths or hungers or simple inattentions have made it hard for us to see what's all around, including those pale sight-lines to deep time and the enormous expanse of the past.

And to wonder what good it would finally do if we could. But surely the long view is better than the myopic one. The former reveals patterns, the long undulations of rise-and-fall waves, the cycles of famine and feast, warm and cold. Though it might induce a maddening passivity ("Just wait. This too will change"), it might also reveal the web of life in some more loosely woven, long-ago state, its inhabitants with more room, more freedom from constant threats of survival, a state when a creature's inherent shape-shifting possibilities had time and space to emerge and that tortoise who lumbered along the path could grow gargantuan without even a dim instinct to fear me.

Some weird propulsion of isolation and mutation has steered even us down our human path. "Humanity," Loren Eiseley says, "is as much the product of a figurative genetic island lost in time as is the giant Galapagos tortoise." Perhaps that's what we, some millions of years ago, needed too: time and space with a little break from the constancy of threat and fear to develop opposable thumbs and bigger heads. I don't know. I don't know enough about human evolution or what experiments failed, left some early form of our species straggling in the savannahs. But I like to think that's what those interior islands of my childhood were all about: refuges without pressure, experimental labs, a chance to become what I, without them, might not have, whatever that might be.

* * *

Deciding not to pass that stumpy-footed tortoise on the path, I sat down behind it and watched it lug its heavy shell toward whatever food it could find. I knew even then that on those isolated islands my sense of time, and therefore the world, had expanded. What I worried about was how short-lived that expansion might be. What would the trip become after I returned home and faced whatever new horrors—shootings, train

43

wrecks, the buffoonery of certain politicos—were sure to yank me back into the all-human, all-critical, *now*?

I didn't want the place to feel like a dream, a magical world that one must travel far to visit, a memory that cannot exist well alongside the frenetic pace at which most of us live most of the time. How to hang on to the expanded world and the self who believes she can thrive on occasional islands where time becomes something more than a way to structure a day? I knew that back home I'd be able to pore over photographs, re-read my notes, cling to phone calls with my traveling companions. Remember the iguana holes? I might ask. And the cave with a dead sea lion? But I doubted that reminiscing would be enough to equal how pricked alive we and the past all felt. Or to keep us from marooning ourselves in current time with its compelling crises, its calls for action *now*. How easy and inevitable it is to backslide into the isolating, stranded-in-the-present atoll of *today*.

Ahead of me, as the tortoise lumbered on, the glinting sun lit up a clump of lichen on its back. I had not known lichen could grow on a shell like that, had not been able to imagine those wispy pale-green gardens fuzzing the backs of creatures slowly creeping across lava fields. What better image than this of the altered sense of time in the Galapagos.

In spite of the urgencies of day-to-day drama that can focus the mind, such images can help us feel it again, that ancient history rubbing up against the present day, the glimpses that can meld, at least for a moment, the faraway and up-close worlds again: dried mud of the past, lizard-birds and small dragons among highways and cell towers. Remember this, I told myself, sitting still on the path, watching the back of the tortoise slowly recede; remember the "faint shining" of the Galapagos doesn't have to be just a one-time glimpse into the past at all. Maybe it can be a door any of us can—with a little work--imagine ourselves back through and thus forward to a sense of the world as deeper, larger, older, more in flux and therefore maybe a little truer than we've ever known.

Lamentation

3

*If only there were time. But the north is now melting, the south's too
hot, the east is flooding and the west's on fire, and it's tempting to aim
up again, to appeal to the old-time gods. Looking for lifelines to higher
realms once soothed when gloom was too great and miracles like
parting the sea and walking on water seemed to promise quick rescue
and escape.*

*Here at the edge, though, gull cries sound like protests, sand crabs
scuttle thoughts of elsewhere, and the response to the question*

what does one do when one's desire / Is too difficult to tell
from despair?

*sounds something like this: Listen to wise neighbors. Become more
than one. Become two or three, twenty, forty, become the congregation
hunkered down in the midst of what is not a miracle at all—no parting
of water or walking on top--but the ordinary way of the world as each
glinting wave, each crafty drop of a gull-carried clam, every precise
stab in the sand by a foraging plover pin us firmly to the marvels of
this world, whose diminishment we think only we can grieve.*

Immersions

As horizons vanish and perspective narrows, the up-close, in-your-face world can seem like the only world there is. When I lifted my head, I could see the shore--bit of the sandy St. Thomas coast--but mostly I kept my head down (am I avoiding something?), breast stroking above sea grasses set to swaying by waves and tides whose rhythmic advance and withdrawal keep the food moving, the underwater forests teeming with birth and decay while who knows what more is brewing in this veiled, half-hidden world.

For the short while I was swimming above the reef, there was only this shrunken, twilight world of waving lace and soft grasses. Gone was the research that brought me here: What's climate change doing to underwater worlds? Gone was the country's history of colonialism and plunder and the aftermaths of hurricanes. There was, it seemed, only the water, the wavy vegetation—600 million years old--the soft strands brushing up against my skin.

I can find no proof that Monet ever stepped into his pond at Giverny. We know that as the years went by and his eyesight faded that he moved his easel closer and closer to the pond. But did he ever roll up his pants legs and wade right in? Study the pond from its wet center? Probably not, though surely he longed to immerse himself in the water-lilied world, to be--like so many artists--absorbed by his own subject.

Or perhaps that's exactly what he did—immersed himself, yes, but as an artist does—with one eye on perspective. It's true that by the time Monet died there were no borders to his paintings, no suggestion of pond's edge or lawn. But Monet pulled off what I think must be one of a human's greatest achievements: to be intimately immersed in one's obsessions while continuing to widen the frame, to expand the larger vision and see not just the individual stalk of a lily plant but the whole habitat, niche, and biome. Or in Monet's case, to see in the lily and the water the light all around, vault of sky and reflection of clouds.

What if our new stories had different shapes and movement? Maybe narrative does something other than arc from A to B or, following Freytag's Pyramid, climb to climax and drop to resolution? Maybe it moves instead more radially from the center out or like a pulsing jellyfish in a series of contractions and expansion. Or perhaps it accumulates like barnacles on a shipwreck—by accretion and layers. What if perspective emerges from a series of collaged vignettes, a deepening polyphony of haunting impressions, a mosaic of layered patterns?

Perspective is what I did not have during those days in the Caribbean. The world was up close and intimate, luscious and mesmerizing, but I had, while I was there, no ability to see as an artist or scientist might. I knew enough—that no sharks swam nearby and that water lilies don't grow in the shallow saltwater off the coast but that big beds of algae do. Some float on the surface, their long strands dangling like untied ribbons into the watery forest below. Others rise from the ocean floor like cobras.

But without perspective, I couldn't see then their history and function: for millions of years, seaweeds have formed these tangled habitats. Their watery-green, dim-lit havens just under the surface form the foundation of near-shore and tidal homes. Their proliferation helps filter light from above, prevents the kind of brightness that can damage tissue, subdues the rush of incoming tides, hosts the daily drama of hide and seek as thousands of invertebrates slink among stalks, slither under a wet mat of algae. Without a deeper sense of time, it's hard also to see the recent trend: as oceans warm, some seaweeds thrive and others don't. Some kelp like it hot, some don't.

Some seaweed forests have already begun to drift poleward. Fleeing the warming water, they are taking with them the fish, mollusks, countless invertebrates that hide and dwell among them.

But species unable to release their holdfasts are doomed to stay in one place. Theirs is a story that might soon dwindle to nothing. Imagine the demise: an underwater heat wave carried by currents begins to slosh among the seaweed beds, bathing delicate stalks. This is no luxurious, spa-like treatment. Rising temperatures can stress a plant and cause cellular damage, triggering a reallocation of its resources.

Forced to "choose," some plants will struggle to repair their damaged cells before they grow bigger or reproduce. If the damage is severe enough, seaweeds will grow listless, less able to resist the onslaught of pathogens which act like underwater flames thinning out the forests. Diseased—bleached--, the seaweeds will grow thin and sparse, allowing more sunlight to penetrate shallow waters, illuminating the underwater jungles and their myriad shelled, gelatinous, clawed and tendrilled creatures, many of which need shadow and murk to grow.

More warmth stimulates more bacteria and on and on it goes, the increased brightness and warmth undoing what has swayed mostly out of sight, quiescent and rich for millions of years.

Among all the experiments and slow adaptations, what remains is the ruthless drive to survive. Acclimating takes energy. So do adapting and moving. The effects of ensuing re-arrangements will cascade throughout the webs of communities both off-shore and on.

They may cascade through the mind, too. Monet might not have changed the world per se, but he changed forever how we see the flux of water and light, the mysteries of reflection. And changes like that—the growing ability to have perspective on what we're immersed in—is one way to avoid the mono-maniacal attachment to any one thing that can lead to fundamentalist stories, territoriality, possessiveness, xenophobia, an inability to adapt to whatever lies beyond the home range.

It's an ability that might also trigger other challenges: among the shifting images of ourselves and the communities we live in, can we see ourselves as creatures in flux too? That is, can we imagine less destructive ways of living?

Can we put aside a theology full of miracles and god-heroes who rule by thunderbolt and decrees, old epics that pit us against a natural world they tell us has been broken ever since some woman ate an apple?

There are stories so deeply embedded in our cultural consciousness we don't even know they're stories.

Those seaweeds—and the turtles, the coral, the myriad beings--have no myths of wrecked gardens of Eden, no chances or reasons for redemption, no afterlife to strive for. They have—as do we--this one planet only and its deeply entwined, complexly webbed strata of bedrock and life.

Removing my flippers, I waded slowly ashore and headed for a nearby snack bar, aware of how quickly I could cross the sand and then the lobby, cool tiles under bare feet, and order juice from an orange I wasn't even sure grows on that island. Sometimes, I confess, I barely know where to begin.

Meanwhile, at some point, those heat-fleeing seaweeds will find themselves stranded at the cooler edges of islands. In Australia, they're already congregating on the southern edge of the continent. That will be their end. Beyond that edge, there's no place to go.

First Extinction | end of Ordovician Period
443 million years ago

Over the next 50 million years, a second round of experiments cranked up, slowly at first, and then with increasing speed. By this time, the monster continent known as Gondwana was sliding toward the South Pole, and much of the northern hemisphere was tropical, covered by a shallow ocean that sloshed against the shores of continents that look nothing like they do today. Warm and shallow, the sea resembled a world-wide kiddy pool in which some bits of marine algae drifted.

And then a few tiny organisms grew a few tiny tentacles, then a shell, a stalk. There were primitive corals and a lot of drifting, floating changes and then the mutations, the experiments, the *try-this, try-that* trials ratcheted up, kept churning, small bacterial life mutating, niches multiplying, plant and animal life diversifying, a whole multi-stratified array of new life-forms both in the water and edging on to land. Geologists call this "Diversity's Big Bang."

Eventually the earliest creatures with a skeletal rod appeared. With gills and tails but minus anything like a jaw, they roamed the bottom of those shallow seas sucking food with their down-pointed mouths and paving the way for the later evolution of sharks.

And for 100 million years this new life—primitive, tentative--eked out another beginning, a quiet one among more dramatic events—occasional meteor debris and huge volcanic eruptions. Elsewhere, the increase in exposed rock from towering volcanoes and the uplifted Appalachians led to more silicate weathering and eventual erosion, which drew carbon down fast into those shallow waters in the form of biomass which died and sank.

Plummeting levels of carbon means plummeting temperatures and soon the Gondwana continent stalled at the sun-deprived South Pole and started to cool. Glaciers began to churn their inexorable way up from the pole, sea levels dropped and the air cooled fast and all the while the hundreds of new species just getting a foothold in what had been a balmy, hospitable world suddenly had to deal with the stress of an inhospitable ice box. Most couldn't do it.

Success rates for this second round of experiments hovered around 15%. In this first extinction linked to climate change, most everything else vanished.

And thus another boom, another bust.

Attachments

Much farther north, along a rocky edge of the North Atlantic Ocean, a mess of mussels at my feet might have appeared to be motionless, but it was restless with hunger and the urge to stay alive.

It's an urge, of course, that drives us all, makes many of us sink our roots deeper, hunker down more determinedly.

A mussel digs in with its foot that eventually secretes a kind of glue that twines into byssal threads—a bit thicker than human hair--which grip the surface. Picture a home anchored by short ropes on a cliff-side.

As the tides move up, mussels open their bivalves, filter seawater for plankton and zoospores, close them against the intrusion of a hungry whelk's proboscis. More mussels congregate, sidle up, attach themselves with foot and byssal threads in protective beds to the rocks and one another.

That is the ancient way. In story terms, if plot is the sequence of actions you've always undertaken in order, say, to stay alive, then you keep doing what you've always done. Double your allegiance to what's worked in the past.

What happens, though, when the ancient plot stays static but the setting is suddenly, radically altered? When the increasingly acidified ocean waves slosh around and under the mussel beds?

This: acid eats away at those threads, changing proteins in the silky glue, weakening the mussels' attachments.

Seawater's rising, and more violent storms hammer the coasts, pound the mussels whose more fragile threads, stretched by the force, sometimes snap. The mussel is flung who-knows where, increasing the force on those remaining, which are now more exposed. The living blanket is ripped, and on and on it goes until not just the mussels are at heightened risk but so too are the communities of crabs and snails and worms that live within the blanket and the birds that feed here, and on and on that threat, too, goes.

And here's what's next: Across the planet, CO_2, pumped into the atmosphere for decades by the burning of fossil fuels, sinks into the oceans and reduces the carbonate ions available for shell-building. When the thousands of mussel larvae that have been swarming over and among the mussel beds are ready to settle down and grow a shell—as they must--they may find ingredients in short supply. Protection, then, may be thinner, more fragile. Porcelain walls instead of concrete.

Easier to peck, if you're a gull. If you're a whelk, easier to bore a hole into.

If you're a thin-shelled mussel on the verge of being yanked from the safety of your bed, then that foot that once saved you might be about to undermine your chances of survival.

If only mussels could do what some other threatened species can: disconnect, change the setting, migrate to better habitats. For mussels, though, there's no escaping the acid, and, once attached, they can't move anyway. Their kind is 245 million years old. Staying put is in their DNA. Theirs is a story, one among millions, that requires its ancient setting. And once that setting is gone, their story may be over. Thus they remain at the mercy of us, a species capable of mercy but often too preoccupied with ourselves to be so inclined.

And so when a storm moves in, they cope by inching themselves around in their crowded beds, shifting on those ropy attachments that grow brittle and stiff. Like moored boats swinging on old anchor lines, they aim their narrow ends into the storm and stay together, as if hanging on until the fury has passed could alter what happens next.

In the end, attached mussels cannot go elsewhere.

Lamentation

4

Neither, finally, can we, though our kind, thinking we can, has started fast-walking backwards as seawater seeps into streets, into our shoes while the mud of floods, still vibrant with red worms and pea crabs and clumped with soggy scrapbooks and picture frames, lies caked on curbsides to be scraped off by bulldozers and hosed down the drains.

Here at the shore, where the buildings are being up-lifted on stilts and the seaward road stops at the new sea-stopping wall, a can-do part of me turns back and sets about the business of erecting barriers and coastal armor, mitigates, repairs, survives,

while the orphaned part of me lingers, presses an ear against the wall.

What happens if the un-real begins to feel real?

On the other side, the water sounds as water always sounds when it's slapping against a wall.

We move on anyway towards what's almost gone

and hope, having turned its back on the ancient and elemental, grows unreliable now, and misleading.

Exchanges

On a crescent beach between rocky points, the sand looks still, un-urgent, the essence of a Zen garden, though no monk is in sight and whatever rake might have combed patterns in sand has disappeared too. The sand, however, is anything but motionless. Like a grainy behemoth, it even "breathes." Incoming tides force air out of sand, replacing it with water; outgoing tides suck water out, allowing air to flow back in. The rhythmic exchange leaves evidence in the sand: pits and nail holes, tiny volcanic cones, blisters I could pop with a finger. Whole Lilliputian worlds of tiny doors, wee apertures, invitations I could imagine accepting because I'm interested in what can happen if we linger in small openings and feel not just a disorienting kick to consoling notions of solidity but a welcoming sense of roominess underground.

Under its surface, sand is a massive swaddle of larvae and eggs, squirming cradle of the newly hatched. To mole crabs, home must seem like a three-dimensional maze, some passageways leading up toward sunlight and food, others aiming down, toward deeper, cooler sand-below-sand. They're hungry, clamorous. Searching for sustenance and mates, they scuttle through underground slivers of passageways opening and closing between and among the grains as wind and water and the footsteps of human and gull press from above.

They aren't alone and the rest aren't still, either. When the tide comes in, millions of other creatures scuttle to get out of the way of tiny underground avalanches, to form new corridors. They scale bits of sand, swing themselves along strings of grains, glide and undulate, sink hooked mouth parts into bits of grit, scramble to keep the tunnels open to food sources and mates.

The further into a hole I go or imagine, the bigger the world seems to become. Once there was solace in the assumption that most things adjust, find a new balance, settle down. But as the seas rise and keep rising, those under-sand tunnels will soon be water-filled sluices that never dry out. If that monk were to return to the now-altered garden, he might sit for hours in the silence. I don't know what he'd emerge with: an escape plan? a koan? An image of emptiness?

Lamentation

5

*Or maybe a close-up film with the action speeded up, that same monk
sprinting, the rake jittering in his blurred, moving-too-fast-to-see hands.
The sand sprays, furrows empty; tiny ridges erupt, sand critters race
up beach and down, grow bigger gills, smaller claws, stronger byssal
threads. It's all frantic dash and lickety-split and happening in a second-
-geologically speaking-- because the water's rising faster than anyone
predicted, hastening coastal erosion, squeezing beaches between higher
sea levels and humans' approach from inland and storms ravage not just
by wave action but by increased sediment washed into rivers, which dump
their load as they enter the ocean, fouling the sandy edges.*

*And if the speeded-up long view shows us the time for tranquility is over,
and pummels us with proof that what happens on the hillsides affects what
happens to the rivers and the rivers to the coast, the coast to the sea and
on and on it goes, then how are we, standing here and now, how are we to
revise our stories, our notions of self?*

Death Throes

One morning, determined to not look away, I watched a sea star in disease-triggered death throes. The grisly business begins with white lesions on the body. An arm shrivels; the suction power in its remaining feet starts to wane; the body bulges in odd places. Beach-combing kids poked at it with the toes of their sneakers, and I didn't stop them. Had they ever even seen orange ones glistening in tidepools? Ever marveled at how they re-grow a snapped-off limb? Observed, even felt, their own ideas about regeneration suddenly expand? Have I?

For 450 million years, healthy sea stars have been voracious predators, keeping other species from taking over tidal zones. Sick stars, however, mean too many mussels, less room on the rocks for others, decimated seaweed beds, diminished opportunities for the systems to thrive.

Marine biologists with microscopes scrutinize the residue for clues to the infection perhaps by a pathogen gone haywire in warming ocean waters. They call the disease "wasting syndrome." I'm not sure what to call this intentional witnessing-- a choice I can still make because my own body is still intact? An exercise in futility? A playbill from a theater of the grotesque?

Seventy percent of the planet's sandy beaches are now in retreat. Whatever design might have been in place for eons is being obliterated by stampede, the present Anthropocene rushing headlong into the future, ready or not.

When its internal organs rupture, the sea star near my feet that day will finally explode. Gooey remnants will wash ashore. What kind of new story can help us live with such an accusation?

Second Extinction | end of Devonian Period
364 million years ago

The next round? Fish, fish, and more fish. Big ones, armored ones, jawless and big-jawed, lobe-finned and ray-finned. And reefs—an explosive proliferation of reefs and the communities they support: algae and sponges.

And Placoderms—thirty-foot-long ocean-prowling monsters with heavy armor and protruding head-bones that acted like swords.

And fish-like creatures attempting to move from life in the sea to life on land. Out of the water, they tentatively scooted and flopped. Ready now with something resembling skeletons and muscles, they found they could lift their heads enough to keep them out of the mud, overcome the hardships of gravity, thin air, threats of dehydration to become—eons later—amphibians and tree shrews and what eventually became us.

Oxygen levels in the atmosphere reached almost 21%, close to current levels. Lungs inflated, nostrils opened and closed.
Meanwhile, a whole barrage of shallow-water, tidal-zone plants headed onto drier land. Tangles of creeping roots wriggled into crumbly rock; branches grew skyward. Tall, weed-like trees spread inland. Wingless insects proliferated.

Back in the salty waters, the slow recovery from the previous extinction inched ahead and then halted.

One theory: as the tendrilled roots of that inland expansion of vegetation snaked their way into tiny fissures in surface rock, those tendrils became small roots and then sturdy roots, and they pushed downward with increasing force, Stone began to crumble, building soil—good for trees--and releasing long-held minerals, especially phosphorus, which eventually washed into the sea where algae gulped the nutrient-rich concoction, bloomed wildly, and then, dying, sucked the oxygen out of the water.

Anoxia—dead zones in the ocean—multiplied. Marine life began to choke, reefs to shrivel. Jawless fish were asphyxiated. Sponges disappeared.

Meanwhile, up on the newly greening land, all those trees experimenting with uprightness were gobbling so much carbon from the atmosphere that CO_2 levels dropped by more than 90%. Temperatures plummeted again. Erupting volcanoes caused spikes of mercury in the atmosphere.

This was not an overnight obliteration but two waves of decimation that lasted for millions of years. By the time it was over, 75% of all species—mostly marine—had been forever wiped out.

For all their dominance, even the giant, heavily armored placoderms could not survive dying oceans and the onset of a deep freeze. For them—and for so many others--this was the irreversible end of the story.

Blooming

Deep freezes some 15,000 years ago formed much the New England landscapes we know today, including the former-glacier-now-lake in which I learned to swim so many years ago. Standing on its shore recently, I remembered only fragments: the gradual wading in, the immersion, feet disappearing, calves, thighs, a slow-motion vanishing into murky water thick with seaweed and small fish. Swimming has more recently felt like a chance to re-position the body, to move parallel with--not perpendicular to--the curved surface of the planet. Some kind of righting of a too-upright, spiritually ambitious posture.

Years after I left that lake, I heard that algae blooms had invaded it. Before I knew what that meant, I pictured a swath of millions of tiny blossoms undulating like water lilies on the surface of the lake. *Bloom* meant a flowering; it meant a stage of vigor and health. I pictured a pointillist painting of a lake, a park, big trees and parasols, the scene idyllic and robust and me still swimming there, learning in some literal and figurative way that to be immersed means not just to lower the body, but also to be involved, to be *in and among*.

Some algae trail filaments. Others sport tiny vesicles that make them buoyant, able to rise to the surface and soak up a little sun or descend a bit into darker, denser, cooler waters. They drift around as single cells or tiny strings or miniscule colonies. They're attracted to warmth and sun and sometimes when conditions are just right—phosphorus and nitrogen and water temperatures--they congregate on a lake's surface and suddenly reproduce explosively, billions of cells replicating and crowding and spreading. These blooms can be so large that orbiting satellites above them depict them as massive chiffon scarves swirling on the water.

A recent bloom in Lake Okeechobee covered 33 square miles.

Large-scale blooms off the California coast recently made the national news.

And along the southwest Florida shore a bloom recently stretched for 150 miles.

But nothing about a real algae bloom is idyllic or robust.

Seen from below, from underwater, the blooms resemble dark ceilings. They block sunlight, shade the water, gunk up fish gills. Before long, down in the now-murkier depths, aquatic plants—accustomed to filtered sunlight—struggle to photosynthesize and begin to collapse in fibrous heaps on the muddy bottom. There they join the bodies of those fish.

The riotous bloom itself can be short-lived, meaning that when it, too, begins to decay, it piles up as dead matter that fires up bacterial life, which gobbles the oxygen that aquatic fauna need to survive until the lake, dark now and depleted, is on its way to being what biologists call a "dead zone." A tomb.

Excess nutrients from agricultural run-off can cause a bloom.

So can a warming planet.

Average air temperatures have been steadily rising over the last few decades; ocean temperatures are increasing too. But it's the world's lakes that are heating fastest. Researchers, in fact, have recorded increases close to a half-degree per decade, which might not sound dramatic, but to a system as finely tuned as a lake's, such increases can cause havoc.

And if the blooming algae is the 2.3 billion year-old blue-green type—*cyanobacterial*--the bad news keeps on coming. Certain species release toxins that can poison shellfish, kill cattle and dogs that lap at lakes' edges. An algae bloom in Lake Erie a few years ago so polluted the drinking water supply that more than half a million residents of Toledo, Ohio, kept their faucets turned off while they lined up to buy bottled water.

One species irritates the skin, resulting in what many freshwater swimmers know as "swimmer's itch." Other species, through contact or ingestion, can damage livers and kidneys, cause incoherent speech, even respiratory paralysis.

Though we can work to reduce nitrogen and phosphorus in our rivers and lakes, no cool-down of the planet is likely, meaning the algae will continue to flourish. Is there a remedy? One might be more human ingenuity--sieves to filter the water, scum busters to scoop up the blooms, neutralizers to change the chemistry, ultrasonic technology that blasts the lake and creates a sound barrier on the surface that algae, trying to float up from below, cannot penetrate. Without sunlight, those inventors claim, the algae will die before they bloom.

Another remedy, of course, might be more human restraint—less burning of fossil fuels, less dumping into lakes--and more awareness that stories of high-tech wizardry can keep us in the dark, well insulated from convulsions of an unstable world.

Long before I was a child who learned to swim in that lake, the river that drained it was dammed for the Saugus Iron Works. Before that, the lake teemed with spawning salmon and alewife. Wampanoag tribes taught their children stories of respect for the land. Caribou roamed the hills and shores. Before that, mastodons thumped across the land, drank of that lake, and giant beavers built their lodges there. And before that, long before that, the lake was a shallow bowl that held a chunk of ice 250 acres big and a half mile high.

And long, long ago, before the ice and the mastodons and the dams and warming planet and that girl who swam in that lake, long before all of that, billions of years ago, that blue-green algae started it all. Through some unknown combination of who knows what, ancient Cyanobacteria in the Archaen and Proterozoic Eras began to photosynthesize, producing the first traces of oxygen that allowed, eons later, for life as we know it to finally evolve –tree shrews, early lemurs, monkey-apes, and finally the primates who stood up straight and evolved into humans as we know ourselves today.

And the algae has been here ever since--persistent, resilient, heat-seeking, phosphorus-and-nitrogen-loving blooming stuff that started the whole photosynthesis process and resulted in oxygen that animals could breathe and on and on things evolved from there.

And now that we're dumping more stuff in our lakes and turning up the planet's heat, it's blooming more often again, more profusely. After I left the lake where nobody swims anymore, I began to swim in Chesapeake Bay whose temperatures have been rising more than a degree a decade. There's even an acronym now: HAB—Harmful Algal Blooms—and standardized warnings posted along shorelines: *Warning: Harmful Algae Present.* The urge to immerse myself there vanished. Upright, I've walked along the edges of coves and heeded the signs.

I swim laps in a pool these days. Back and forth from one chlorinated end to the other, I try to remember the puny body I once was, immersing myself in what's left of a 13,000 year-old glacier and learning to question such givens as gravity, the usual laws of locomotion, and the rightness of always being upright. Even now, I still pretend that swimming is a small but daily way of repositioning myself as a minor character in a much larger story, of being part of some Ecclesiastical pattern. I break the surface, turn my head, fill lungs, empty lungs as the past recedes, the future looms, the planet warms and the planet cools, algae creates, algae destroys. For decades, it has seemed like a worthy rhythm to be part of.

But what to do when the rhythms have gone awry? When a thing on its way to becoming its opposite might take longer than we have? Or might happen in a heartbeat? *Breathe in, breathe out* can calm anxious nerves but we'll need more, of course, than that-- if not *immersion in* then at least a better *alignment with.* Without a major re-positioning of who we think we are in relation to this planet, every turn, turn, turn at the end of a lap these days keeps reminding me Pete Seeger could be wrong, that even he might no longer swear it's not too late.

Lamentation

6

*Some of us know the consolation of grief-rocking, of sitting on the ground,
knees pulled to chest, torso swaying forward and back while the pelvis,
that cradle of uterus, cradle of birth, tilts to the front, then to the rear, as if
some hand holds it underneath and angles it to and fro like a yawing bowl
of soup, unable to decide whether to spill it, and if it does over which bowl-
brim—front or back--the overflow should go, whether the real power of a
lament is forward, like a warning, or backward, like a tribute*

*and some of us sway in this primal ritual of comfort as we do with our
babies, until finally we know that the cradling hand is not our own, that
this is no hemming and hawing about which lip to over-brim, that this
rocking is, in fact, the ancient, pendulating way of the world, the great
moon-pulled sea sloshing to one side of the earth and then the other, as we
with hugged knees lean first toward the future, then toward the past, grief
swagging back and forth across the body's core until what's been lulled
is not just sorrow but also the fear of total spillage, of being emptied, of
having nothing left through which to re-enter the world.*

Refuge

When I have felt emptied and alone, left with nothing in hand with which to re-enter the world, my mind has often gone to stories of refuge. The ones I grew up with come first: Jesus as savior, God's love as a haven, the sanctum of heaven. But none of those stories is set where I am set—in the central Appalachians not far from a small river, in 2020, on the cusp of global catastrophes. They're stories that could be told in Turkey or Rwanda, today or two thousand years ago. That, I suppose, is what makes them timeless and therefore available to us all.

But more and more, those stories seem rootless, free-floating above the worlds of lichen and trout lilies, tick-pests and pollution, isolated from the pressures we struggle with today. Doesn't it matter that the stories we try to live by take place in worlds we actually live in?

The stories that come next to me are Buddhist. Trying to find a way back in, I feel the ancient sway between the old idea of "refuge" as safe harbor and the notion that to "take refuge" is to stay doctrinally homeless. The former means protection (from what?); the latter means doing without the safety of protection (to what end?).

Conundrums like this make my head spin, which is one reason I'd been outside for hours, contentedly following Savage River upstream under limestone outcroppings and through the tangled slicks of rhododendron when I strolled into a grove of ghost-like hemlock trees. Tall and regal, they should've been flush with fresh growth. Instead, their branches were bare and at their feet lay the tell-tale silver-green litter of dead needles.

If I had better eyes, I might have seen the culprit—a bug that hitched a ride some sixty years ago on a transport from Japan. From their arrival in the southeast, the wooly adelgid have made their way up the central Appalachians and are now slowly infesting their way north. Sinking their piercing mouthparts into the base of one hemlock needle after another, they settle in to feast, sucking the sap, growing stronger while the needle weakens and drops, depleted, to the forest floor.

Without needles, infested hemlocks cannot make sugar out of sunlight. Photosynthesis stalls, and the trees—some of them 200 years old and a hundred feet high—grow skeletal, wraith-like. Their shallow roots grow frail. Eventually, they topple.

Meanwhile the adelgids, buoyed by milder winters, spread north into forests once too cool to sustain them. All up and down the Appalachians now, they bask in the earth's warming air, mating more often, hatching earlier. Think of incubators with the dials turned up.

Ghost trees are no good at blocking the sun. Here and there along that river, the hemlocks have thinned, the canopy has opened, and now sun streams into what used to be cool and shady pools.

This is not good news for the brook trout, who have lived in clear cold waters for millions of years. Sun warms the waters and decreases the oxygen, which means the fish can suffocate in their ancient habitat.

I don't know how a fish feels change, whether it registers alarm, how it knows to head farther and farther upstream where it searches out deep pools, undercut banks, fuller canopy, any place where the adelgid hasn't yet decimated the awning that healthy hemlocks provide. But I want to believe that if any creature could tell the story of its own extinction, it wouldn't leave out the surrounding context. In the trout's case: the hemlocks, the bugs, the way we humans went on wrecking its habitat.

As streams keep warming, the fish keep retreating to cooler waters closer to the source. Though their species name *fontinalis* means "living in springs," at a certain point, the streams aren't deep enough. Nothing big can swim there. Their bellies would scrape the bottom. Caught between too hot and too shallow, whole populations can languish and die.

Brook trout are beautiful. Shimmery olive green with gold speckles and bellies that glow pale orange, they look like underwater mosaics. They know nothing, of course, about "taking refuge" in uncertainties or in the comfort of old stories, nor would they call those vanishing cold-water streams and deep shaded pools a thermal refuge in a warming world.

But we, who might, often find refuge in action: we convince the forest managers to introduce the British Columbian beetles who prey on adlegids. Or to re-plant stream banks with red spruce. Or to get rid of the pest; re-build the habitat. We humans find comfort in solving the problem, doing *something*, though there's no refuge in thinking anything is fixed. Impermanence *is*.

Which makes it easy to understand, then, why we might give up, spend our days wandering along rivers, refuse to even imagine the mostly inaudible sounds: steady slurp of sap being sucked and the faint ping of hemlock drop, belly of a brook trout scraping in too shallow waters. Tell me again: Why try to save a thing?

Lamentation

7

Because when a brook trout dies, it isn't just the end of one moon-speckled fish that's moved back upstream, but also the momentary collapse of a small river-born community, the reduction of ripple and sun-glint that means for kingfishers and herons more hours of stalk and dive

and when a chipmunk dies, what's emptied too is every burrow it once inhabited, every hard-packed tunnel sinewing below forest floor and every image it once saw or imagined in whatever tiny brain its tiny skull once sheltered and one swath of acorn-sated flush that might have radiated through the woodlands as a sense of abundance

and when a lactating deer dies, so likely will the fawns hiding on spindly legs just a few yards into the forested side of the road, while the suck-away sound of final breaths is so almost-audible we think we can hear it just before the rapid-fire series of subtractions starts and we're left with less, less, less, and no talk of death's inevitability can erase the fact that right there in the water, the forest, the woods is another and another giving-back to nothingness

until the weasel slinks in and vultures swoop and the owl and their babies too, just learning to nibble and munch, and the maggots and worms as the world re-shrugs and, all around the pocked and torn display of death, re-asserts itself, gathers, accumulates, re-absorbs until at some point in the long night of one grief or another, that old belief that death is a discrete and isolated moment with an endpoint

is beaten as if into grist, finally stripped of language, meaning stripped of meaning, as we remember those human-bodies-become-ashes in ancient fires into which, if we weren't wailing, we might stick a finger, draw the hieroglyphics that tell of those shifting states of being, untie the knotted dead-ends, maybe learn to live as what we are—not just alterable parts of the lie that death is singular and unaccompanied-- but also ingestible, morphable bits—everyone of us--of life everlasting, while still and still knowing that even that is not nearly enough. Amen.

Dead-End Hosts

Ah, the despair of never enough. It creeps, as the insidious often does, so slowly you might think it's doing no harm. Certain creatures are like that, too, including one that can live along river banks, like the one I stood on one spring in south-central Florida. A few feet away, the Caloosahatchee River ran warm and meandering, the kind of river I like to swim in. But the locals say no, don't do it. The water is full of blue-green algae and bull sharks.

The grass beneath my feet, though, might have been full of something else--fire ants, stink bugs, and a species of snail I want no part of. Deaf, mute, nearly blind, indifferent even to its own offspring, fond of burying itself in soft ground during the day to avoid a hungry rat or the subtropical sun, the Giant African land snail is big and slimy, the kind of creature I might feature if I were a maker of cheap horror shows.

A loner, it can, in theory, even fertilize its own eggs. In reality, its hermaphroditism is mostly ineffective and so it does what most of us do: mates with another. Coupling is a gooey affair that begins with tentative tentacle rubbing and tilted shells, a little biting around each other's genital pores. After they've coupled and fertilized each other's eggs, they ooze apart to resume their solitary lives. A week or two later, they'll lay a clutch of some 200 eggs. At five or six clutches a year, that's over a thousand offspring a year.

The US Department of Agriculture has declared the Giant African land snail an invasive species. Voracious, it feeds on over 500 kinds of plants, gnaws on carcasses, even sand, especially if available plants are low in calcium.

But it's also an invaded species. Outwardly, the snail may live alone; inwardly, it has company. In fact, by the time it's an adult, thousands of larvae of a tiny worm have likely moved in. Or, more accurately, have been ingested when the snail, gliding over infested rat feces, bit down and swallowed.

Once inside, the larvae settle into the snail's muscle tissue and molt twice into 3rd stage larvae. An intermediate host without consciousness, the snail, of course, would make a lousy movie villain. It can't be accused of insidiousness. There's nothing duplicit about its behavior, nothing intentionally harmful. It knows nothing of what we, shrinking from images of parasitic infestations, might call a horror show. Laden but not harmed by the larvae it hosts, it continues its gluttonous foraging, wiping out bananas, peanuts, cauliflower, beets.

Meanwhile, the larvae inside bide their time until the snail is devoured by something larger—often a rat. Once inside, the larvae wait until they've reached the intestines of the rat, at which point they break through the intestinal wall, enter the bloodstream, and head for the brain where the larvae mature into their almost-final form. Now almost an inch long, the creatures then slip back into the bloodstream, which carries them to the right ventricle of the rat's heart. It's here in that tiny chamber that the worms reach full adult status—transparent with a tapered end, complete with ovaries or testes and two or three triangular teeth. In the week or so they spend in the heart, they mate with other congregating worms and lay their eggs. The eggs, pumped out by the heart, slip into the bloodstream which carries them first to the rat's lungs, where they hatch. Lungs are good for hatching, evidently, but not for inhabiting, and so the worms bust out of the lungs and wiggle into the trachea.

If this were a movie, I'd slow down the action here, move the camera in, up close, and film the swarm, up to thirty of them, emerging in slow motion from the lungs like a many-tentacled monster.

The rat, perhaps sensing for the first time that something is amiss, coughs. Up come the worms, which the rat then swallows and, sometime later, excretes.

And then along comes the Giant African land snail who lowers its lipless mouth into the pile of rat feces.

And so the cycle continues. The larvae that had initially been excreted by rats are eaten by snails, which are then eaten by rats. It's all rather elegant, a closed circuit of host and excrete that has likely been on-going for thousands of years.

Climate change, however, has already begun to darken this picture, too. As temperatures continue to warm, the Giant African land snail with its parasitic cargo has begun to range northward. Since its arrival in this country aboard a barge in New Orleans around 1980, rat lungworm larvae have now been found in California, Alabama, northern Louisiana, along the Gulf Coast, and even, to scientists' astonishment, in Oklahoma. Their expanding range has, according to a recent study, "changed dramatically."

Not only is the lungworm ranging into more heavily populated areas, but it's also invaded at least three native Florida snails: Florida amber snail, perforate dome snail, and the quick gloss snail. "Alarming," another researcher says.

And of course if this is some low-budget horror movie, we humans would have to do something more than contribute to the warming that's expanding the snails' habitat. We'd have to be part of some more immediate dread. And we are, because rats aren't the only creatures who dine on snails.

We humans eat some of them deliberately. We also eat some of them inadvertently--tiny snails or their residue left on lettuce, cauliflower, beets, and bananas. We don't always scrub our spinach. We ingest the tiny larvae which set out to do what lungworm larvae do: make their way to the host's brain. A woman in Hawaii describes the sensation: "The parasites are in the lining of my brain" she says, "moving around."

But when it comes to the worms, here's what distinguishes us from rats: if the worm larvae make it to our brains, they're finished. For reasons biologists don't understand, the larvae get stuck there. In the rat brain they move on, travel to the rat's lungs, reproduce, squirt out its arteries and into the lungs where they're coughed up, swallowed, excreted so that the meandering snail, gliding by on its glistening foot can eat and begin the process all over again.

Not so with us. The larvae, stranded in the human brain, can go nowhere. The worm, unable to complete its life cycle, is doomed to linger in our heads. Infested, we can do nothing but wait it out. It can take years. It feels, the woman says, like having "a hot iron inside my brain."

In the worm's world, that makes us the "dead-end hosts." If you're a worm who sets up shop with us, that's the end of you. We accommodate but we don't nurture. We house, we're afflicted, but we allow no exit.

Now officially invasive and illegal, the size of a human fist, the Giant African land snail with its load of parasitic larvae can't be called insidious. Oozing northward across lawns, through backyard gardens, along riverbanks like that one along the Caloosahatchee, it's merely taking advantage of a growing range and doing what it's done for millions of years just to survive.

The same cannot be said of us. What creeps, what's creepy, what may make us dead-end hosts in much more global ways, is the effects of our unwillingness to see how the stories so many of us westerners have told ourselves have split us from this world. Part of me wants to hear all this as mere metaphor. Part of me knows this is the real horror story.

Third Extinction | end of Permian Period
250 million years ago

And then came millions of years of rebuilding again, rebounding, species evolving and spreading, oceans full of sharks, rays, sponges, and sea-lilies, shores crisscrossed by slithering reptiles, dry land dotted by early trees, the gymnosperms and cycads which provided, for the first time ever, a little shade for the pre-mammals that began to trudge across the land. A pig-like creature with a turtle-like beak appeared, along with insects that developed mouthparts for piercing and sucking, beetles, cicadas, lizard-like things with primitive sails on their backs.

The life force seemed well on its way again, experimenting with this or that, new creatures evolving, some disappearing, the usual slow-motion shifts in shape and habits that mark a healthy evolutionary process. Competition between two groups of animals was underway: the Synapsids sported one hole in their skulls and the Sauropsids sported two. The former were ancestors of mammals; the latter ancestors of reptiles.

If anything with historical consciousness had been alive back then, it might have looked around at all the flitting and crawling and stampeding, all the evolving and adapting, tryouts and trials, and felt relieved that the planet, having overcome the devastation of earlier mass extinctions, seemed flush again, churning along in relative tranquility.

But around 250 million years ago, evidence suggests, underneath some remote part of what is now Siberia, the earth began to rumble. Humongous volcanoes erupted again and again, burying two million square miles of land with red-hot lava two and a half miles deep. The lava flowed on top of the land and it flowed through fissures underneath where ancient layers of coal and gas had been deposited eons earlier. The lava ignited the layers, spewing greenhouse gasses into the atmosphere. Levels of CO^2, one of the planet's great climate regulators, went haywire.

Over the next 60,000 years, a mere tick in the geologic time clock, multiple systems collapsed. So much released carbon means a massive sudden warming, more acidic oceans, anoxic oceans, plant-killing acid rain, even an increase in deadly hydrogen sulfide spewing up from underlying rocks.

Nothing much can survive a deadly concoction like that.

That pig-like creature Lystrosaurus hung on; perhaps its burrowing habit spared it. Plant life was poisoned, animals suffocated, oceans stagnated. In the end, the extinction known as "The Great Dying" was catastrophic. Ninety % of marine life was extinguished. Seventy % of land vertebrates were obliterated.

The cataclysm came perilously close to being the final chapter for any kind of life on earth.

Epilogue

It goes without saying that California redwoods are immense. They are, in fact, the tallest living things in the world, and they do everything tall things should do: they tower, they reach, they bring to mind words like *lofty, exalted*. As they soar upwards into a shimmer of light and clouds, some of us might even think *sublime*. I certainly have, every time I've been among them in Muir Woods or farther north along the California coast.

But these days, I confess, such looking disappoints. It's not that I'm tired of grandeur or that neck-craning gets harder as I age. It's more that whatever the real questions are-- *what does it mean to accept our own deaths? the extinction of species? or the end of life as we've known it?*--I've come to trust I can ask them better if I'm looking in less obvious, less elevated places, searching not overhead but underfoot. Maybe that's because, as writer Robert Macfarlane says, the below-ground world "still hides so much from us, even in our age of hyper-visibility and ultra-scrutiny. Just a few inches of soil is enough to keep startling secrets, hold astonishing cargo."

For several weeks, Robert Macfarlane's book *Underland* kept my gaze lowered. Chapter after chapter, he lured me to underground labs, catacombs, buried rivers, and caves. For many nights, I dreamed of fungi and burrowing owls, tunnels and sinkholes. For days, my morning walks in the woods took on another dimension, the whole forest expanding downward and then opening into an underground immensity just as far-flung as the galaxy overhead. I kicked a lot of stones, crumbled bits of these ancient Appalachians, all of which made me rethink a recent afternoon in Pennsylvania where, map in hand, I had driven three times up and down a curvy half-mile stretch on a back country road just north of Harrisburg. I was looking for a pull-off site, a

"State Forest" sign, some kind of trail that would lead me to the single, foot-high bush that's been growing in one particular spot since before the time some of those redwoods were mere seedlings--over a thousand years ago. For months, I'd been in a mood to consider mortality, including of course my own, trying to weigh the lure of longevity against the power of limits.

And then Macfarlane's book arrived, and I began to understand the connections I had overlooked.

* * *

Once I found the unmarked, pull-off site on the side of the road, I parked the car and stuffed cell phone and notebook in a pocket. I'd wanted to imagine I'd be on some kind of solitary quest through a tangled valley, ducking under outcroppings, checking and re-checking my map, a modern day Ponce de Leon hunting, if not a fountain, at least an example of long-life. So what that it's not the next miracle drug I sought but a bush that's managed to stay alive for a thousand years?

But, though there was solitude, I felt little sense of quest and no sense of achievement. The path, short and well-marked, took me right to it. The plant is a relict, almost eliminated during the last Ice Age. Somehow, a few pockets survived the brutal cold, including the patch I knelt in, which, botanists estimate, is at least 1300 years old.

The box huckleberry—*Gaylussacia brachycera*, also known as the Bibleberry—looks like any number of low-growing, oval-leafed plants, like teaberry or blueberry. Botanists think the ten-acre site in Pennsylvania consists of a single strain of protoplasm spreading everywhere it can—up the base of a tree like an evergreen collar, around pine saplings, into the trail itself. Pennsylvania's only other huckleberry colony—also a single plant—has been growing a few miles away for ten thousand years.

Few other living things continue for so long to operate under the natural world's overarching mandate to keep spreading the DNA. Not much in my husband's body and none in mine can comply with that directive. Nothing we do from here on out will contribute to gene pools or natural selection or to the care and feeding of the young. His sperm have grown sluggish. My egg production slowed to a standstill years ago. Evolution is not interested in us anymore.

Still, on a personal level, there are the rewards of continued good work, family warmth, the next good game, the pleasure of seeing of Motherwell painting or a black bear ambling along the edge of our field, along with making contributions to the cultures we value--art, music, literature--and to the causes that matter.

But on a purely biological level, there is on the horizon for us—and for millions of others--only the passing of time, increasing senescence, and general decrepitude—one system after another failing at ever-increasing rates. How, then, can we, do we, ought we, think about our own mortality?

In *Deadeye Dick*, Kurt Vonnegut writes, "If a person survives an ordinary span of sixty years or more, there is every chance that his or her life as a shapely story has ended and all that remains to be experienced is epilogue. Life is not over, but the story is."

Epilogue, Vonnegut suggests, is shapeless afterthought, post-vital-élan deadweight, the tedium of unproductive dwindling.

For years, I resisted this grim pronouncement and all that it suggests of finalities. Though I don't believe in any heavenly life after death, I've wanted to imagine ways our lives persist, to recognize, for example, that our story is always being absorbed into a different story. That its previous shapeliness might be on its way to a different shape. Or that it has dipped below ground to become part of the soil from which emerges whatever's next.

But does such thinking, I wonder, serve mostly to soften the final blow and allow us to side-step that most un-western notion of the value of limits? When the poet William Blake declared, "The road of excess leads to the palace of wisdom," I suspect he could not have imagined the excess to which some might go for something close to everlasting life. In California's Silicon Valley, numerous bio-tech companies spending billions of dollars to reverse the aging process are aiming for human lifespans of 120, 150, 170 years. One executive decries this society's "pro-ageing trance" and claims we can push for 1000. In western countries you can pay $80,000 to have a surgeon sever your dead head and freeze it, in hopes that geneticists will use your DNA to grow a spanking new body so you can have it reattached and carry on another hundred years.

Wisdom? Even Blake might have reminded us of the lesser-known second half of his famous quote: "for we never know what is enough until we know what is more than enough." Somewhere between too much and too little is where we learn sufficiency and how limits help us to distill and refine, to concentrate and shape. But these are rational thoughts, which seldom matter when death or extinction is imminent, when the longing to go on and on stirs and thrums in the about-to-be-buried body. The underland,

Macfarlane warns, is 'terrain with which we daily reckon and by which we are daily shaped." It is the realm from which both limitless sorrow and hunger swirl.

* * *

My husband has Parkinson's Disease. He's a poet whose days are still full of work and play. He writes and reads, back-hand slams a ping-pong ball past his opponent's astonished face, makes love, listens to music, visits with old friends. Parkinson's is incurable and degenerative; his movements are often halted by episodes of freezing. Some days he can't open a bottle of wine. There will likely come a time when he has trouble swallowing, when he can no longer walk at all, and we will be faced with questions of feeding tubes and assisted living. Below those looming dilemmas lies a central question: how far will we go? For the sake of living longer, how far will we venture into the world of more drugs, new clinical trials, extended physical therapies along with likely incontinence, wheelchairs, and bed lifts? If there's a choice, it will, of course, be his, not mine. Sometimes we talk about this. Sometimes I go for walks in the woods.

* * *

With age, supposedly, comes wisdom and here's the paradox: the ethical question we should be ready to ask in old age might jeopardize our chances of living to an even older age, that is, just because we're a wealthy country that *can* pour billions of dollars into research that might assure that our failing bodies won't fail just yet, *should* we? I'm not talking here about preventing or curing diseases that shorten the lives of the young or ways of improving the general health of a population or about anything we can do to ease someone's suffering. What I worry about is the mindless goal of doing whatever it takes so that we, the mostly privileged, can push longevity into our eighties and nineties, even hundreds. I suspect many of us in western societies live as if our futures were, Vonnegut be damned, shapeable.

I worry, too, about how medical funds are distributed. I don't know if spending billions of dollars for research into artificial knees means spending billions less for malaria or polio vaccines, AIDS drugs or cleaner water. And I don't know how to calculate other costs of longer lives: more assisted living facilities, more Medicaid and Medicare, more people around for more and more decades consuming more and more in a world with increasingly limited resources.

All of us chafe against limits that incarcerate, especially ones imposed by others. But maybe mortality is a limit that can work the way all *good* limits can. The work of the artist--and all the rest of us—is, finally, about creating within frames, boundaries, restraints. We can rail against death, pour money into cryonics or colonies on Mars, or we can feel mortality as a container that might help us fashion wiser, more tempered, more exquisitely formed responses to one another, to our homes, and to this planet.

How to balance those questions when I'm the one watching my husband try to get from the kitchen to his study with a cup of coffee in his shaky hand? Of course we opted for the complicated procedure known as deep brain stimulation. Medicare paid something close to $90,000 for it, and for now that coffee stays mostly in the cup.

* * *

It turns out that the story of Ponce de Leon's search for a Fountain of Youth was pure fabrication. The truth is that a historian named Gonzalo Fernández de Oviedo y Valdés wished only to discredit Ponce, and so reported to the Spanish court that the gullible explorer was off in the New World following cockamamie directions of some jokester Native Americans toward a fountain everyone knew did not exist. Ponce, in fact, did no such thing, dying instead at age 47 from an arrow wound in Cuba. But the fountain story was a good one and even though people knew even then that such a quest was silly, designed only to make Ponce looks foolish, it's the story that's persisted and become another historical lie. Such is our romance with eternal youth and with heroes who reject Vonnegut's claim that *Life is not over, but the story is.*

In the meantime, my husband contemplates a future suicide, the possibility of a decision against prolonging the kind of life he doesn't want. It'd be sensible, he says, meaning he's thinking, as Vonnegut does, of his story and an ending he'd like to write himself. That's his kind of frame. Neither of us knows what that might mean.

* * *

A redwood survives by thickening its bark and upping its production of tannic acid. The bark insulates from fire; the acid protects from rot. Both mean a tree can survive in big display in the open air, 350 feet tall, taking up some 38,000 cubic feet of above-ground vertical space.

The box huckleberry does the opposite, and it can live far longer. It survives by insulating most of its actively growing parts safely underground. For someone like me, whose imagination is so enlivened by subterranean spaces that I wrote a whole book on caves, this seems like a good plan. Keep the tender sprouts out of sight, nestled in damp and protected places. The huckleberry's plan—can we even call it a plan?--for growth is, of course, more literal. Some time way back in its evolutionary past, the original stem turned sideways and burrowed down. For hundreds of years, that same buried stem has been pushing horizontally in all directions, branching and re-branching below the surface. It reproduces by cloning, meaning it grows by extending its very same self on and on and on. It doesn't evolve; it doesn't change. In some ways, it's an example of stasis without end, not the kind of life my husband—or I or most people I know—would choose.

Am I suggesting here that Vonnegut might be right? That longevity dooms one to living long after the story really ends?

Reading Macfarlane complicated that notion. His book, in fact, sent me back to that hillside and plunked me down on the dirt where questions of limits vs. longevity dissolve into images of communal intimacies. I knelt in the center of a low-growing plant that had been there for eons and pushed my fingers into the ground, feeling around for anything resembling a single stem or rhizome. There was none. Even the gentlest prying-up revealed a snarled mat of fibers and soil. Dirt got under my nails. As I brushed hair from my eyes, small clods smeared my forehead. Though a plant has no idea what it means to live a singular, storied life, the huckleberry reminded me there's more than one way to be vital. Below the surface, there was no untangling, no separating. No end to intertwining and connections, the intricacies of a living colony. The realization nudged me to try to broaden my definition of "alive."

In spite of its emphasis on deep time and long perspectives, *Underland* is ultimately about "Nature . . . as an assemblage of entanglements of which we are messily part." It's about familiarities and closeness and immersions, about going down into, feeling the earth, the cave, tunnel, catacombs rub up against your skin. As the book makes clear, the stories of deep time cannot help but arrive slimed with mud, drenched with seawater, or coated with sand. They come from below-ground places in the earth; they carry bits of the planet's grit. To get at those secrets, that intimacy, you have to get a little dirty. You have to get down to the layer where things are buried.

* * *

If my beloved dies before I do, any deep-time perspective I might have will likely implode for a while and I will, like most mourners, be left with the gone sweet grubbiness of two lives twined together for so long, the intimacy of sorrow, and the effort it takes to make it through another day.

I can hope that grief will ease later and I might recall the bigger picture. To remember the long history of the world, its changes big and small, is to be reminded that no matter how well or ill-shaped or how limited or how *over* our own stories are, they are always—I still want to maintain--embedded in larger ones. We do this and that today. We get the job, find a lost friend, accomplish whatever. The coffee spills or it doesn't. Meanwhile, continents still drift, mountains wear down while others rise. I might be talking about grandchildren here as well as the enduring power of art, waters from uplifted mountains, seeds careening across glacier-scarred landscapes, huckleberry bushes that have survived since way before Jesus promised everlasting life.

How to think about the end of all of *that*? About the mortality of our entire species, the million other species now heading toward extinction, the habitats, communities, whole biomes? Does Vonnegut's claim apply here too: *just enough story, not too much epilogue*?

In our hubris, we've exceeded our limits, sullied the waters and filled the atmosphere with gasses now choking our planet. "We stand," Macfarlane says, "with our toes, as well as our heels, on a brink."

Perhaps, as some claim, our story as a species is almost done and all that remains is an unwanted intimacy with grief on a scale no private loss can prepare us for. In our disregard for limits, we've sacrificed, it seems, the ability to control what happens next and what legacies we'll leave. And so perhaps the Anthropocene is, in fact, all epilogue now, and our obligation is to stay with the trouble we've caused, see it out to its dwindling end.

Or maybe our obligation is to devise a different kind of epilogue. The notion of story or a life as linear is, in itself, a limitation, the type to disregard. What if we saw epilogue as a web of sprawling mutualities spreading everywhere it can, radially shaped, even kaleidoscopic, branching and re-branching below the surface, underlandish entanglements of mud and green and sea and of the human and more-than-human worlds in which we, chorus-like, lose and find ourselves again and again.

Not the deadweight of aftermath but hints that a new story is already underway—how to live with grief, how to provide hospice for whole species.

The quiet suggestion of some kind of sequel.

The underlands, Macfarlane says, remind us that what will outlast our species is not love but radioactive waste and plastic. But then he offers us the image of an open hand that he's threaded throughout his book and the scene with which he ends: he and his son are out in the woods; he's kneeling on the ground as the boy raises his hand. "I reach," Macfarlane writes, "my hand towards his and meet it palm to palm, finger to finger, his skin strange as stone against mine."

A gesture at the intersection of intimacy and abyss, which is, finally and urgently, where we now have the most to learn.

Lamentation

8

When I read about drought and you read about fire, when we track the sea rise in Alaska, when you study glacier melts and we note changes in bloom times and I chart the dying reefs, when the experts hunker down with piles of data and dire warnings, project calamities, design imperfect solutions,

how weary of fear we become, how tired of thinking until we take to walking in shaded valleys and from time to time look up at the zillion stars and down at the moon-lit river and across the hills turning coral then drab, and forget, for once, to comfort each other with talk of how long the earth's been spinning and, instead, stand together--I, you, we— mute, in the stellar dark

Temporary Homelands

Maybe it's a growing despair that made me want not just to see but to *smell* the arrival of spring. Some kind of primitive sense-reminder that, in spite of it all, another season had indeed come around again. Standing by a small vernal pool in Connecticut, I, like a Thoreau character after winter, wanted to sniff the ground, get a good whiff of earthy moisture and tender growth.

Vernal pools are now-you-see-them, now-you-don't affairs, deep-in-the-woods, concrete demonstrations of transience and of the ancient truth that *everything passes, nothing stays*. Shallow ponds with no above-ground outlet, they appear in early spring all over the world, slowly filling with rain, transforming muddy, leaf-littered depressions into small mirror-like pools, reflective circles, invitations to throngs of frogs and salamanders and seasonal musings by the likes of me. And then they disappear in mid-summer, dry up, grow silent and devoid of small creatures. Their rhythm of empty-fill-empty is so precariously linked to rainfall and warmth you'd think no creature intent on survival would ever select one for a home.

Vernal pools trigger big questions about temporary homelands and how to live here on this planet with an increasing sense of impermanence. But the real draw, for me, at least that morning, was much more mundane: I wished to smell the muck and shallow water rimmed now with reminders of resurgence--an array of new grasses and greening shrubs--and an even stronger wish to glimpse the pool's springtime host of returning critters, especially the tiny fairy shrimp that spends its entire brief life in a home that's almost equally brief.

The shrimp's story—like the story about ourselves that some of us have been drum-beating for the last couple of decades--is the story of a race against time. Only its race is not the kind of global pressure of an eons-old planet warming so fast we might all be doomed. Theirs, in fact, is the story of how to successfully live a whole short life before the shallow water that is their only home evaporates.

Originally inhabitants of ocean waters, fairy shrimp are tiny crustaceans. At some point during their 500 million years on this planet, the ancient dance of predator-prey forced the fairy shrimp out of the seas and into lakes and then small ponds and finally into vernal pools where shrimp-eating fish cannot go. Less than an inch long, they sport feathered appendages, which they use to cruise—usually on their backs--around their shallow pool. Their bodies are so thin they seem almost translucent. In their race against time, they molt and molt again over a couple of weeks, add appendages and stalked eyes, change color from orangey-bronze to greenish-blue.

The urgency? In just a short while, when mid-summer temperatures warm the pool and oxygen levels drop, the water will begin to evaporate. The pool will shrink and grow more crowded. Whatever's left of their life cycle has to happen fast.

But the fairy shrimp is ready. Within a few weeks of her own hatching, she's molted again and filter-fed, scraped algae from the bottom, dodged predator beetles and waterfowl. And she's mated. So all that's left, then, is to drop her fertilized eggs to the muddy bottom before the pool dries up. Job done, she dies soon after.

By early June, the brown-gray dampened ground will resemble a dirty bathtub ring around what's left of the water, whose level will have already begun to drop. Soon the bottom will be exposed--soft mud and then baked mud--and then winter will freeze the whole shallow bowl. If I were walking there in January, I'd have no idea what kind of soppy incubator, frog-croaking hootenanny, shrill chorus of tree frogs and legions of back-swimming fairy shrimp the place had been home to and will be again. Crossing from one side to the other, my boots would crunch—the only sound--across frozen leaves.

Meanwhile, down in the mud those eggs—cyst-like and well protected— would have been programmed to hatch at wildly different times. This is the fairy shrimp's way of hedging her bets. Though the cysts need the dry mud of winter, newly-hatched shrimp need the rainfall of spring. If the rain doesn't come and the pool doesn't fill, a few cysts might risk hatching but a good portion will remain tightly sealed, wait it out for a wetter, better, year. Some will wait for decades.

Given the iffiness of sufficient water, most other critters might give up on the place, abandon the pool, crawl on their stubby legs to somewhere else in the woods with a more reliable supply. The fairy shrimp, however, cannot leave. It has no wings, no legs, no belly on which to squirm to a better place.

Which is why where vernal pools are doomed, the fairy shrimp is doomed.

It's not just because of bulldozers, roads, and shopping malls. Sudden floods can wash out a vernal pool or fill it with silt. Agricultural run-off can increase salinity, reduce acidity. It's not easy for a vernal pool to flush out poisons or regulate its temperatures. Picture dumping toxins into a plugged-up bathtub, not a free-flowing stream that at least has a chance to rinse itself clean. Drought dries the pools; extended drought destroys them. On the other side of the country, 90% of vernal pools in California have permanently disappeared.

Is this the point at which we too-often re-chant that ancient refrain-- *Everything passes, nothing stays*?

Of course there will be other pools, other fairy shrimp. But what's to stop us from extending that serene *laisse faire* attitude to other, now clearly temporary, homelands—like wetlands and clean rivers, coastal regions, the Arctic, the whole warming planet? Though 90% destruction means that millions of fairy shrimp, suddenly and permanently homeless, have forever perished, those tiny crustaceans are obviously not among the greatest of my worries. They are, however, one harbinger among many, reminders that even the canniest ways of hedging our bets for the future won't matter one whit once the homeland is going down for good.

There by that Connecticut pool, the fairy shrimp I'd hoped to see were staying out of sight. Maybe I was just too late. Maybe the shrimp there had already died, their ethereal bodies floating like torn fingernails above the flattened nest of blackened leaves, the sopping, matted bottom already growing dry. A whole host of eggs lay banked, nevertheless, and waiting for another season. I was tempted to kneel there, push my fingers down into the mud, feel around for the still-dormant cysts, finger them, even, as if they were rosary beads for the non-Catholic, a reminder not to count prayers or beg for favors but to practice another ancient tenet: to try to see the world—maybe even smell it?--as best we can, without delusions, before it—including us—vanishes.

Lamentation

9

*And to try to say it, as best we can, before the tongue grows too brittle and
stiff to say not just what's irrevocably dead but also what's uncertainly
missing, the ambiguously here and not here, a new lament whose first
job will be to step in and voice the grief that a brain trying to process
trauma—individual and global--cannot so easily feel, a lament that will
replace ugly words—*broadband *and* optimal, virtual *and* networked *with
resurrected ones—*distended *and* gorgeous, mossy *and* webbed.

*And to say them aloud, sing them aloud, hum them and hear the way to
say the word conjures not only color and texture but also place—that edge
of the field where light dances and the palate shifts under sun and clouds
and the time of day—until the tongue and eye are so linked to the land
they create a place for the body there too and here, along river banks,
among the glens and pastureland and within—within!—the folds under
outcroppings, flesh on stone on grass, on knoll and vale.*

*How sterile can a language become before the tongue grows numb? Before
it cannot infuse the mind and hearts of our progeny with tonal descriptions,
before it cannot say to the future, Listen, this is what and how we grieved?*

Empty Dens

Rambling in the early March woods in central Appalachia—ground smudged by last fall's leaves and the twiggy debris of an already-ending, too-mild winter—I saw the black mound of black bear ambling fifty feet away. She had big face like a dog's with a giant snout and a stomach I couldn't see but which had to be—after four or five months without food—shrunken, those inner folds dry, like corrugated cardboard, rasping against themselves. Two cubs scrambled behind her, looking all the while like those stuffed bears my children used to sleep with. The big bear raked the ground, stuck her nose into a leaf pile.

Her presence out and about this early meant that somewhere in these folded hills between Piney Run and Finzel Swamp, there was now an empty den, a hollowed space beneath a fallen hemlock or in the lee of a wind-blow or under a small outcropping. The den had been likely dug out by claw and snout, kept clean of feces by her fecal plug. Sometime two months ago those cubs had been born and the sheltered den had been full of breathing and the scramble of tiny paws as the cubs jostled up the soft belly where they rooted and kicked their way to the pink nubby source of the milk from their mother who would've moaned and rolled and lifted her woozy head to see what she had done.

Once, on expedition with the Department of Natural Resources, I hiked for miles through snow-crusted woods and helped pull three bear cubs out of their den. As the official folks tested the mother they'd tranquilized, I nestled the infants inside my jacket to keep them warm. Whimpering, they tried to scramble up my chest, poke their heads out the neck-hole of my jacket.

I remember their wet-dog smell, sharp claws, their blue eyes, widened, perhaps, by having been hauled above ground weeks before they were ready. I remember the mother bear, paralyzed, nearby. I remember the sheer joy of cuddling the cubs and the cringing guilt: I knew I was being watched.

After they'd weighed and tagged the cubs and checked the mom, the biologists re-positioned her back in the den, and I, nestling the cubs, rubbed Vick's vapor-rub on their fur to mask my human smell. One by one, the biologists took the cubs, crawled into where their mother lay, and snuggled them in close.

By all accounts, they resumed their semi-hibernation, dozing and suckling, until the weather warmed, skunk cabbage emerged from the winter ground, and they all staggered out to an early feast of tender green shoots.

That was years ago.

Those bears I saw the other day should not have been out yet. For another month the cubs should have filled that den with the sounds of mewl and suck while the mother snored until outside the hills greened-up and young plants pushed through the thawing ground, turned a winter landscape into a feast of ready-to-eat food. At some point soon after, they would have roused themselves and stumbled out, as if arriving at a banquet just as the dishes were being spread out on long tables.

Instead, she, up on her groggy feet, had already emerged, lurched around, half-awake, to find that the same warmth that roused her had not yet warmed the soil enough to send the first green shoots up through the ground. The cubs had stumbled out, too, watched their mother rummage in the half-frozen dirt for skunk cabbage roots, pluck catkins of aspen, paw a few early grasses and the sparse supply of over-wintered berries.

What had so discombobulated that bear's timing, caused her to show up while the ground was still mostly frozen and food was just a future plan? Maybe a poor diet last fall when drought could have dried up the usual autumn bounty? Or, more likely now, a string of early warm days, premature triggers for den emergence.

"Phenology," scientists call it—the study of seasonal cycles of hibernation, migration, bloom dates, all the exquisitely-timed occurrences that mean just as robin eggs hatch, insects do too and that worms emerge just when the plants they need begin to leaf out. And if those insects are also pollinators, then flowers bloom in time to take advantage of the nectar-seeking bugs' ability to transport pollen on their legs, meaning the flowers will reproduce and the meadows and woods will be bursting with bloom just when the herbivores and bees need them the most. It's all so orderly. You can even buy phenology calendars, phenology wheels, phenology charts, all designed to help you track what to expect when.

If the timing is right, then nourishment is available, one course after another, all up and down the food chain. And if the timing is wrong, and flowers, for example, bloom weeks earlier than they used to, then we get marching bands in Cherry Blossom Parades trampling the withered blooms that have already dropped and littered the streets and sidewalks.

Climate change throws the whole system out of whack. For the critters and plants who rely on physical surroundings and cues like temperatures and hours of daylight, the timing is less and less predictable. Bloom times that Thoreau noted in his journals some hundred and fifty years ago happen now weeks earlier.

* * *

Imagine reading a thousand-page book whose binding suddenly deteriorates. We'd been halfway through it. It seemed the plot—about human survival during some kind of catastrophe--had been carefully worked out and we're eager to know what happens next. But then a door opens and a breeze scatters the loose pages, which are not numbered. To put them back in the original order is impossible. The traditional transitions of a story-- *first, then, next*—only make us think there *should be* order. There's not. Not these days. Not with the kinds of stories that are and are not working now. We can't even know what chapter we're in—the opening? The climax? The epilogue? Our job, in fact, is not to unscramble the original story but to make a new one.

This is the task, now, here, in times like these with constant news of floods and droughts and bears out rummaging before there's any food to eat. We will have to learn to live with non-linear plots, disrupted progressions. We will need to create alternative ways of linking one event to another or find ways of accepting disorienting narrative gaps and multiple simultaneities rather than careful sequences. We'll need new definitions of resolution. Imagine our lives depend on it. Imagine a billion other lives do too.

* * *

The synchronization of birth and budburst, leaf-out and bloom time all evolved over many generations and now has to adjust. And a lot of it does, though often at different rates, meaning some creatures starve, some don't and so we get caterpillars emerging weeks before birds who rely on them have started to migrate north. We get bears up in the hills scrounging for meager supplies while down in the valleys we humans keep filling the birdfeeders as if it's still winter, keep putting our garbage out on the curb at night. A bear's ability to smell is seven times stronger than a dog's. They'll go where they know the food is.

We get those prematurely emptied dens, reminders perhaps of other homes that had to be abandoned and of other inhabitants—humans, for instance--whose hungers have sent them crossing boundaries, roaming into territories where they're treated like invaders.

Lamentation

10

*At first the end was a relief. Winter over everywhere, the redbud and bluets
bloomed. Forests greened and the hidden life below ground rose up through
burrows, up tender stems and sapwood, out into the warming days, while the
life that the cold had vanquished flew back north from the South. So what if
it was still January? We opened the windows, breathed in the thawing dirt,
watched robins stab the newly green grass. What's wrong with a premature
spring if it lifted the spirits after the dreary dark of winter? At the Spring
Equinox we lit bonfires, harvested St. John's wort, verbena, and rue,
transferred summer rites to spring, pushed the spring ones back to winter.*

*Soon, we didn't bother unpacking heavy clothes. A couple of cold days
we layered up, lit a fire, told one another the cold wouldn't last and it
didn't; it warmed right away and everything we'd known about cold
and hibernation, the interior days of long-ago winters, became the stuff
of legends. It's better, we said, to be warm. Isn't this more like the good
fortune of Eden? How like us to want the end to feel like
another beginning.*

*As the real end neared, rivers dried, hayfields withered; the relentless sea
rose inland. We cancelled bonfires and avoided the sun, told ourselves it
didn't matter that we could no longer live where we used to live. We'd dig
underground vaults to store our seeds, huddle in bunkers built to shelter
us from flames.*

*In every fairy tale, in the beginning ends as a story of consequence.
Slouching toward the future, we remind ourselves now that promise
was once a relief.*

Fourth Extinction | end of Triassic Period
206 million years ago

What is it that cranks the trials back up again? What relentless push squirms for another round? And so it was that after the Great Dying some 50 million years earlier, life began, not to re-surge exactly, but to hobble back, especially at the poles, where the heat of the Permian hadn't scorched almost everything alive.

And then the rains came. Deluges and torrents, millennium after millennium, until the one giant continent Pangaea that had formed eons ago was swamped and out of the swamp, the reptiles emerged. The Triassic was the age of the crocodile, the dominant land predator with longer hind legs that allowed it to belly-run on land.

Small dinosaurs, pterosaurs, frogs, and salamanders appeared, along with mammal-like reptiles that could, for the first time, eat and breathe simultaneously. The first turtles emerged, too new to have shells, but with thick and sturdy ribs on the verge of fusing. Lizard-like, they clomped around freshwater lakes on their stubby legs. Small mammals evolved, mostly nocturnal and solitary, some with more advanced ears and bigger brain cavities. Some began to stand upright for short periods of time.

Out in the hot oceans, coral, for the first time, coupled up with zooxanthellae which turned the expanding reefs into dazzling mounds of greens, reds, and blue. Above them, huge creatures with dolphin-bodies and crocodile snouts cruised, some of them more than 50 feet long.

And there were conodonts, jawless vertebrates that might have resembled eels. With a mouth apparatus of tooth-like projections, bars and plates that eventually littered the ocean floor, they roamed the warm waters.

Meanwhile, under the drippy, sauna-like land mass of Pangaea, cracks were forming. The earth's crust, in constant slow motion, began to split and then to drift in opposite directions. Up from the fissures and rifts, volcanoes bulged and then blew, spewing lava and gasses, upping, once again, the CO_2 levels in the atmosphere which heated and acidified the seas, suffocating the reefs, and killing off close to half of all marine species, including the prolific conodonts whose soft bodies disintegrated and whose mouth apparatus sank into the ocean floor to become, millennia hence, the fossils that would guide oil and gas companies to rich petroleum deposits.

In the end, a mere 20% of species made it through, including those primitive turtles. For the rest, there would be no next.

Uncertainties

If there is a landscape anywhere in the world that suggests the kind of story that's more urgent now, it might be a bog, especially as I begin to hear—out there or inside me?--a rising chorus of lament. As I stood one spring day in a West Virginia wetland, hundreds of miles from the snails of southern Florida, nothing seemed out of place because the place itself seems uninterested in definitions that exclude. There, land and water trade places, and it was hard to know if my foot, which I was about to lower, would hit velvety carpet or a trapdoor to whatever lies below. In places like that, it doesn't much matter that plants are carnivorous, flowers are named for fish, and conifers drop needles every fall the way maples drop leaves. A bog is crammed with oddities, watered by deceptive surfaces. It sabotages categories, erases boundaries, will not allow the either-or thinking that can delude us into believing we know exactly where—or who--we are. Like the stories we need more of, they compel us to look around, sink in, give up the notion of a stable self marching off to subdue the unruly forces of nature— draining swamps, damming rivers, clear-cutting forests.

These are the worlds of hesitancy and qualm, havens for Thomas' Biblical doubts and Hamlet's indecision, the moral morass we want to flee when the next step remains fraught with danger and unseen consequences and we're unsure if there's any ground beneath our feet.

Most of us are not accustomed to the undefined like that. Without solid ground, we want boardwalks and guides, some way of staying balanced while we try to adjust our perceptions and steady ourselves if we must—as sometimes we must--navigate through places between one catastrophe and the next.

Over time, nothing, it turns out, stays the same, not the nearby mountains, the meandering rivers, the ancient redwoods and continental shelves, and clearly not that bog, which, on its way to being something else, is also more than one thing at a time right here, right now. There in Cranesville Bog, my perception of time was widened; the past and future sloshed into the present, became palpable extensions of today. The foreground featured the wet and tendrilled head of a sphagnum plant. But, like the past, a blurred background invited too. Way back, what is the history of extinctions? And what were those dim shapes and what about those color-arcs that can suggest the story of what will happen next? Isn't that one of the enduring questions of our lives—what does the future hold? The salvation stories so many of us have clung to for thousands of years depend on our wish to hear that question answered. But what if that question has now become a way to leap-frog over the present? Riveting stories that catapult us out of our current circumstances are easily more exciting than the stories that slow us down, keep us here in this confusing, angst-ridden time and in these slurpy, sloppy bogs where there is no such thing as an end point.

In a bog, the arc of time collapses, reminding me that below the surface sphagnum, its decades-dead and wadded roots might one day smoke with the old aromas of peat fires and warm hearths. The past and the future coalesce in a bog where it's hard to tease them apart, and so, the bog seems to ask, why not feel their mingling? Why not know the past is still alive and the future's already begun?

The question carries a warning. For these are also worlds of now-you-see-them, now-you-don't shimmer of tannin-black pools and hidden reflections that cannot exclude death. Even plant names remind us: poison sumac and loosestrife, staggerbush, witch hazel, black chokeberry. I plunged my hands in, went wrist-deep into the tangible evidence that what died decades ago can still be touched -- a barely rotten log, alder twigs, even a few sodden leaves from more than an autumn ago, most of the sedges, tiny plant stems, all the woody, fibrous life which lived and died and lies now soaked and still visible below the surface because here's the crux of the matter: debris in a bog decays very slowly.

The microbes that obligingly shred most of the organic dregs of the planet—ingesting, devouring, recycling the residue that would otherwise pile up and stink, leave us over our heads in plant and animal carcasses—don't like to live in bogs. Oxygen is too low, temperatures too cool. The absence of active microbes means what dies here is slow to be re-absorbed into the usual cycle of decomposition, decay, ashes to ashes. Breakdown is stalled, taking centuries to do what in other ecosystems happens efficiently in a few years, maybe even months.

Such a world is made for sphagnum moss. It creeps in from the edges, interweaves its tendrils, begins to cover the surface with thick mats. Its mesh-like roots eventually die but in this decay-stalled environment, they do not disintegrate. Instead, they grow more and more enmeshed, compacted into the fibrous mat we know as peat, a layer I was tempted to stick my whole arm into, to lift off and peer underneath.

What we find here in this 21st century leaves little doubt that the bog's past may soon rise and threaten the future. That's because what was down there, below my feet and out of sight, was not just the languishing rot-debris of centuries but also the billions of tons of carbon those once-living plants—having done what all plants do--absorbed from the atmosphere.

For all that time, the peat has been bog-chilled and the carbon locked in place, inert and buried, held from escaping back into the atmosphere. Carbon, as scientists say, has been sequestered in the bog. They call these places carbon sinks.

Picture a snowball laced with an overdose of air-borne peril that's been pressed and packed and patted hard and stored, harmless, in a freezer for eons. Now imagine what happens when the freezer begins to malfunction, its contents to warm.

Deforestation, over-harvesting for fuel and fertilizer, draining for agriculture—so many ways to wreck a bog. But what's more invisible is more insidious. Global warming is heating those bogs, drying them out, increasing the chances of fire. And as that peat warms and dries and breaks down into humus, it unlocks the carbon, which isn't the only worry. Methane—thirty times more potent than carbon dioxide--is also stored in cool, wet bogs, inert and harmless, until the system, thrown out of whack by temperature changes and human activity, cannot hold it any longer.

Where do all that suddenly released carbon and methane go? Up into the atmosphere where they do what such gasses do—jack up global warming.

And so the drying bogs mean more carbon and more methane escape which heats the planet, desiccating more bogs and on and on the cycle goes until the ancient and languid bog becomes a hot spot of escaping greenhouse gasses. Scientists estimate that a fire can burn half a century of peat and belch tons of methane and carbon into the atmosphere in under five minutes.

The setting for a new story is clear: Three percent of the world is covered with bogs, many with their lids half open now or thinning beyond repair.

Lamentation

11

We used to like the word beseech. *We thought its origins in* visit *and* seek
*could conjure the ancient gods who, we once believed, could supply the
manna or stop the rain, cool the land or heat it up, but all that has passed,
all that has gone, and the history inside us has grown lonely*

*and now the only way to wrap our hearts in other hearts is to move
them—first closer together and then in synch and then, if grief gathers
enough, into the dance that begins* In times such as these *we will remember
lamentation as both anthem and oath, every syllable a small gesture not
just of grief but of refusal—refusal to be silenced, to disappear without a
word into the long barely audible background of loss.*

In times such as these, *we will remember that we too are wedded and
indebted to, begat by, met by, and finally vetted by this one and only earth.*

Between

A few weeks later, a few miles from Cranesville, I watched a bog turtle, looking like a tiny, armored ET, inch along on primitive feet. Evidently feeling safe enough, she extended her head, revealing the orange blotches on the sides of her neck. In spite of the carapace, she's one of the most vulnerable creatures in a bog, threatened not just by habitat loss but by water levels altered by climate change.

Hundreds of thousands of years old and one of the smallest turtles in the United States, her species needs dry hummocks for nests, Having clambered up a grassy mound, she digs, as most turtles do, with her back feet, deposits a clutch of one to six eggs, which she leaves for a couple of months to incubate on their own. If storms pummel the bog while she's gone and water starts to inch up the mound, she'll never know. That job is done; she has other concerns now.

Nestled and dry at the top of the hummock, turtle embryos require oxygen, which enters the porous shell from surrounding soil. But that porosity means other elements can enter, too. If the water keeps rising, reaches the top, pools into the nest-hole the mother has dug, it will saturate the eggs. Who knows what eyes an unborn turtle has, what vision, whether it can see out through the semi-opaque snugness of its shell and watch the rising water all around, on the verge of leaking in?

Should we ask ourselves a similar question? Maybe something about smugness and vision?

The hatchling cannot. And if it's too young, it can't even break its way out. Rising water reduces oxygen in the soil. Inside its own egg, it suffocates or drowns.

The opposite is bad news too. Drought means water levels drop, and though that might keep the eggs safe, it threatens the adults. The bottom of a healthy wetland is riddled with tunnels and holes, squishy places made for hiding. Sensing a predator nearby, the bog turtle's instinct is to disappear, lie still under soft muck until danger has passed.

But drought can hard-pack exposed mud and stiffen soft burrows into cement, leaving the frantically scratching turtle easy prey for the foraging raccoon, the bird with its talons.

If climate change dries out or drenches the bog or swings it back and forth, the turtle can be caught between reproducing and surviving, between dry nests and muddy escapes. For her species to survive, she needs both, and she, unlike us, has no ability to mitigate, no capacity to choose.

Lamentation

12

*Here where small mounds rise matted and damp and the bog cover is
about to give way, there are hummocks we can lean into. What better place
to refuse to think only forward; to refuse the mindless push of progress,
newer models of this and that and bigger-faster too and though it's good
to save lives and work for healthier futures, we cringe in the glare of any
single vision so sleek and bright and spiffied-up it makes us*

*weep even to ask what we're supposed to do with what's left behind, with
landscapes made of matter, of tangled mats and breeding grounds, the
mess and stuff of living every day in the muck and swirl of the earth which
is also the muck and swirl of our most essential lives*

*from which we wish no transcendence, no clean feet, no light-hearted
flitting above the ground out of which we came and into which we'll go.*

*What better place to sprawl over and clutch, to sink into, be enfolded by?
What better place to know that sorrow can make us more supple and
more prepared?*

There for the Taking

How unsatisfying that stories of survival might increasingly depend on acceptance of loss, that enduring might mean repressing the latest leap forward or reducing one's needs.

Think of sundews (*Drosera rotundifolia*) that glisten ruby in a gray-brown, boggy landscape made mostly of bumpy mats. They seem incongruous, arrays of ornate orbs ringed with tentacles that sparkle in the sun.

For us humans, glamor can be merely ostentatious, a doodad-y adornment that masks some deeper absence. But for a sundew, glamour means survival. The sedges and cattails living in nutrient-poor soil have adapted to meager supplies, learned to live, in particular, with low doses of nitrogen. The sundew has adapted too, but not by scaling back its needs, getting by on less. Instead, it long ago changed its diet, forgoing meager offerings from the soil and relying, on a canny ploy of lure and entrap.

Drawn by the plant's ruby color, insects hover close or crawl up the sundew's stem. Unbeknownst to them, the color is speckled with glue and the slightest contact means the insect is stuck. Even a fly glued by a single one of its legs is doomed. It fights, of course, writhes and flaps its wings, tries to pull away. Exhausted, it might take hours to die. Meanwhile, the sundew's clutch response has been triggered. Slowly it folds its tentacles and leaf down and over and around the fly, encloses it in a tightening squeeze and starts to exude an enzyme that dissolves not just the hapless leg but the whole trapped body. Such an inglorious end to the bug; such a shot of nitrogen to the plant. The energy required to trap and digest is evidently worth the energy it takes to keep up its quick-clutching, ruby-red presence in the bog.

Years after Charles Darwin had traveled the world, documenting species and forming the principles of evolution, he confessed in a letter to Charles Lyell, "I care more about *Drosera* than the origin of all the species in the world." It was a temporary obsession with an ancient plant, but while it lasted he wanted to know everything about what it takes to trigger the sundew's clutch response, how fast the leaves fold and unfold, how the plant knows not to waste energy trapping something it can't eat. He experimented for years, dropping light and lighter drops of liquid, gnat's legs, a human hair, recording, astonished, the reliable speed of the plant's fast reaction.

If Darwin were alive, still traveling today, still obsessing about sundews, he'd see something disconcerting happening in Sweden: the sundews' color is fading. They're greening up, blending in, relinquishing their role as glamor-plant of the bog.

Of course he'd want to know why and though he'd figured out so much about the slow evolution of finches' beaks and the barnacle penis, he wouldn't know, couldn't know, that color change is likely the result of recent pollution. The burning of fossil fuels ups the nitrogen load in the atmosphere. When it returns in the form of rain in the bog, the soil collects it, so that now when sundews send their roots into the soil, they find it good— more nitrogen, all of a sudden, meaning less need for meat.

Would Darwin worry about his "beloved" plant's future? He should— though not for the plant itself, which is elsewhere widespread and robust. He would worry, instead, about human activity affecting the slow processes of evolution he'd studied for so many years.

Though they're not likely to go extinct any time soon, those sundews in Sweden grabbing nitrogen from soil instead of from bugs might lose their competitive edge. Outside their carnivorous niche, they're like so many other plants in the bog, rooting around for nourishment. And those other plants have been at it far longer, have evolved the most successful strategies. Does a newly unglamorized plant even stand a chance? If not, how many generations of those sundews will it take for them to know they're out-competed and need to resume their ancient, more reliable ways?

Perhaps the question for us is whether survival might mean reversing some of humans' hard-won advances. And if it does, what kinds of stories will we tell ourselves then? Already, there are green-growth stories, well-being narratives, stewardship stories, alternatives, all of them, to the grand old stories of happily-ever-after or the now-woefully familiar stories of apocalypse. And there are stories told in collective voices, from indigenous traditions, stories of re-wilding, renewal, resurgence and more and more and still the degradation continues.

Maybe what we'll need are less hopeful stories, stories of alliance with the natural world, which, given what's happening, will be stories of humility and grief. Stories that remind us that learning again to say the unsayable can ease the angst of living with illusions. The real test is whether we can make those stories of loss somehow satisfying enough to feel like stories of survival.

A plant can't think about the future. It has only the present and right now those Swedish sundews aren't wasting energy on glamor. Evolving fast, they've cut back glue production, let their color begin to fade because what they need is underfoot and--at least for now--there for the taking.

Lamentation

13

And now finally a different language has begun to uncurl in our bellies; the time is coming when all kinds of things will begin to quicken—the almost living, the sleepy, the oblivious, even the ones who've given up hope. The most serious grievances will go on despite politicians and dreamers, warriors and clergy. But the uncurling language will feel like an earthquake in slow motion; it will sound like a hum from under the dirt. We are not without ambition. Though we believe there exists in each of us a deep yearning to be here, some things must be conjured, then called. Words can act like gods. They can gather us together, make us to stand up. We will not use the words in that way until there is no other way. Today we gather at the edge of the water and keep our hands folded over our bellies. We have a history of patience; we will turn at the right time, lips parted and ready.

Handholds

So maybe the ascetics are right—the way to succeed is with less and then fewer and then finally none. Moss, in fact, reads like a primer on negations: it has no flowers, no seeds, no pollen, fruits, showy blossoms, no fabulous speckles. Absence, however, is not the same as emptiness and moss is no monastery of abstinence.

The gap between two clumps of soil I pulled from Cranesville Bog was choked with drying moss, a tangle of tendril-creep and twine, a dry forest of tiny threads, moss leaves curled inward. But for moss, desiccation isn't death. It's waiting. Some mosses have waited almost half a century, waited for rain, for moisture, for even a drop of wet, waited without waiting, however long it takes.

Part of wisdom is knowing what you are at the mercy of. For moss, rain. For some of us, the mind's restless urge to think ahead, plan for contingencies. Here, as in most gaps, there's a leap to make, yes, something next to imagine and move toward, but it has nothing to do with messages about any afterlife or even the virtues of patience but with the discernment of form, which means, in this case, the shape survival imposes.

When rain begins, moss leaves unfold, uncurl, arch up, spread their tiny leaves, open themselves to the first drops which cling to the moss cell. Neither waxy nor thick, moss—155 million years old--is single-celled and designed for the opposite of sloughing off. Water lingers, soaks in. Moss leaves are shallow bowls, small pockets or slender chutes, little wicks and tubes and pleated plains—all of them engineered for one purpose: to collect rain and hold it as long as they can.

Or maybe the secret is there's no magic in form. Things—including us—take the shape of whatever ups the chance to endure. In this arena of fuzzy architecture, there's no mystery, no bark or woody fiber resembling a secret staircase or hidden basement. I plucked a clump of moss—it lifted easily—and held it like a tangled reminder that might keep me level-headed, save me from heeling over with illusions. Its form's biggest surprise: it's rootless. Attachments? To the earth, it has none. But moss branches are delicately bunched. Their entangling promotes the handholds—in this case, the moisture-holding ones between the plants themselves—that can mean survival.

Fifth Extinction | end of Cretaceous Period
65.5 million years ago

And then, over the course of the next million years, came the dinosaurs. Big ones, little ones, carnivores, herbivores. They flourished in almost every corner of the world—in forests, in what is now Antarctica, along the shores of ancient seas and long-gone lakefronts. For more than 200 million years, they were the key players, dominating almost every terrestrial niche.

Things grew big back then—not just the 7 ton *T. rex*, but creatures of the sea, too—mosasaurs (a carnivorous aquatic lizard 50 feet long, weighing 5 tons)--and of the air--the leathery pterosaurs with wing-spans of 20 feet. It was a world on steroids; experiments in bloated sizes went unchecked. The climate everywhere was relatively warm, conducive. Flowering plants emerged. Competing with one another for the attention of pollinating insects, they exploded into extravagant swaths of color and shapes, a cornucopia for foraging insects and larger beasts.

No wonder the mammals—just beginning to evolve--took it slowly, kept themselves out from under the thundering footfalls by scurrying around at night and hiding in burrows. The earliest mammals—our ancestors—were solitary, rat-like, nocturnal. Their habits might have saved them.

When a meteor slammed into what is now Chicxulub, Mexico, it gouged out a crater 90 miles wide and 12 miles deep, shuddering the earth, and hurling into the atmosphere massive amounts of debris that rained down as storms of hot rocky bits and broiled anything roaming around on the surface. The fires were followed by landslides and earthquakes and a shock wave that traveled for thousands of miles and knocked down whole forests.

Catapulted ash and dust blocked sunlight for months. Plants, needing light for photosynthesis, withered; animals, needing plants for food, starved.

Increased volcanic activity, especially in an area of India known as the Deccan traps, spewed gasses into the atmosphere. It was a time of catastrophic disruption to every water, carbon, and nitrogen cycle on the planet.

When it was all over, 60-80% of all species had been wiped out.

The good news: suddenly—in geologic terms—there were no more lumbering giants casting shadows on the landscape, dominating the food chain, and menacing all kinds of smaller prey. Instead, there were small animals with lower metabolism and less need for oxygen, ones who could scavenge on the abundance of dead carcasses and rotten plant matter, ones who'd survived the cataclysm by hiding underground. Once it was all over, they were the ones, the burrowers-- including the large shrew-like creatures that preceded us--began to poke their heads out of their holes

Haul-Outs

Flying into Ketchikan, Alaska, I studied the gray water and chunks of white ice scattered below. I couldn't tell how big the bergs were—the size of my childhood lake? The neighbor's pond? The ice-speckled world below enticed and, later, up close, eluded me. The problem, I realize now, lies in trying to go toward what's going away. Realizing you can't catch up exacts a price you don't yet know. So does pretending you can. Pressing my forehead against the small plane's window, I watched a whole array that looked like clean, frozen jewels sweeping silently among other frozen jewels that had spilled from a cache high above the inlet.

I had signed up to join a dozen or so other folks interested in species hybrids of Tlingit myths—raven heroes, salmon twins, bear spouses, transformed human beings. Aboard a schooner a few days later, the group sailed up Tracy Arm Fjord where I climbed down the boat's ladder into a kayak and pushed off. Giving a flat berg with a harbor seal wide berth, I paddled slowly out into the fjord when behind me the roar began and I turned to watch a massive chunk of ice break off from the high face of Sawyer Glacier. It was as if the vertical slab had legs and they'd just given way, plunging the ice into the water where it fountained and showered in flinders of sunlit glint. Binoculared folks on the schooner hollered and whooped; one, I was told later, bemoaned the sight as evidence of global warming, and another turned to keep his eye on me. But I had heeded the warnings to keep my distance, and my kayak merely rocked a bit as the resulting wave passed under me, and newly-created bergs undulated and swiveled and then settled into quietly floating chunks that resembled big tables, small stages, frozen, oddly-shaped pancakes.

Skirting a smaller, flat berg with a couple of lolling seals, I paddled closer to a big one. It was huge—the size of a two-car garage. I could see it rocked a little, sparkled blue and white in the gray waters of Alaska. Some of its facets were sheer, almost mirror-like. Others were dimpled and pocked.

I confess it took my breath away. It seemed so majestic, even dignified, as it floated down the fjord, its gazillion crystals reflecting the light, its pinnacle like a bright steeple. If I were much younger, I might have tried to name it--Prism Palace?--and to make up some story about a wizard locked inside and how, as the sun beat down, that wizard would slowly appear beneath the melting layers of ice until the berg became the throne he ruled from and with his wand he could make whatever seemed awry right again. I might have even tried to believe yet another story about concealments and magic, the ancient wish for saviors.

The berg was, I knew, made of freshwater ice that likely began as snowfall in the Stikine Icefield over a hundred, maybe even a thousand, years ago. High in the accumulation zone, a glacier looks relatively smooth and sleek. The farther you travel down the ice highway, the more like a bumpy road it becomes. Although ice bends and stretches, it has its limits. Too much friction over irregular bedrock can cause the surface of the glacier to crack, forming crevasses more than two hundred feet deep. People, sleds, dogs, and snowmobiles can fall in. In Antarctica, the crevasses may be several miles long and a hundred feet deep. In Alaska, some of the most spectacular crevasses occur at the terminus of the glacier, especially if the tongue extends all the way to the sea.

There, the crevasses split the glacier vertically, which destabilizes the whole mammoth face. As sea water splashes up and into the cracks, the glacier begins to calve.

To a kayaker, a calving tidewater glacier can mean danger, but to a pregnant harbor seal, it means safety. When she feels her pup's pending birth, she swims toward what is for her the most protected birthing place she knows: a table iceberg calved from a nearby glacier and now rocking in the middle of the fjord.

* * *

Seals are mammals. Like us, they're warm-blooded, have hair, gestate their pups for about nine months, give live birth, and nurse their babies. Unlike us, they've been around for 15 million years.

Underwater, an adult seal is graceful and fast. Some have been clocked at twelve mph.

Out of the water, they are, like the turtles I watched in the Galapagos, awkward and slow. The one I watched from my kayak that day would have hauled herself up onto the berg in a curious belly-bumping wiggle, a hump-and-scoot action, more like a 180-pound inch worm.

Like mammal babies everywhere, seal pups need to nurse. Fast and often and urgently. They get only a few weeks to build up fat and then they're on their own.

A berg offers refuge. Aquatic predators can still reach her, though a sea lion lurking under water and out of sight still has to get himself up onto the berg itself or knock the berg around enough to send the seal slithering into the water. A pod of killer whales has to coordinate its attack, creating a wave that washes over a flat berg and sweeps the seal off the edge.

Still, for birthing and resting and regulating body temperature, a berg is often a better choice than a pebbly shore or a bit of exposed rock. To haul out there, seals have to time their approach to the tides. Swim ashore at the wrong time and there'll be no exposed beach, no shore to wriggle on to. High tide covers those edges. To wait another six hours for low tide can tax a seal's reserves and put her in danger.

And terrestrial haul-outs increase the chances of disease, mother-pup separation, and the ever-present danger of grizzlies and wolves.

With binoculars, I watched a seal on a berg for a good half hour. She—or was it a he?—rolled on her side a couple of times, lifted her head—was she watching me?—lowered her head, lifted her tail. I couldn't see a pup or any sign of predator. Perhaps she was merely doing what all seals do at some point during the day—warming herself on the ice.

Seals spend a big portion of their lives underwater where they're hard to detect, which makes population counts difficult and projections of populations nearly impossible. Monitoring their habitat is easier. And here's what we know: most calving glaciers are rapidly retreating. When they retreat far enough, they calve less and the bergs that result drop to dry land, where they melt, leaving the fjords increasingly ice-free.

For cruise ships, that might be good news.

For the seals, it's one more temporary homeland permanently gone.

The glacier's retreat can be extreme. Sawyer Glacier has backed up more than a mile and a half in the last fifty years. Over on the Kenai Peninsula, three tidewater glaciers have been in full retreat now for many years.

Northwestern Glacier has receded almost six miles in recent decades. At nearby Exit Glacier, the Park Service in 2019 designated an ice fall hazard zone because giant slabs of ice have been sliding off the face of the glacier, creating ice caves and unstable access.

If the melting were gradual, the seals would figure something else out—give birth more often on terrestrial haul-out sites or follow the disappearing ice farther up the inlets. But the pace is fast and no one knows how quickly the seals can adjust. Already, a court recently ruled that Arctic ringed seals must be included under the Endangered Species Act not because their numbers are low but because their habitat is vanishing. It isn't hard to predict what will happen next.

* * *

Leaving the berg and its lolling seal, I headed back toward the schooner, paddling through what seemed like a debris field of older bergs, all kinds of shapes and sizes—small ones and large. They looked exhausted. Their once-sharp edges had been worn away by water and sun. I watched as the bottom of one rolled up to become a side. Former pinnacles dripped like melting wax; cracks widened. The base of a remnant right in front of me had been so worn by sloshing seawater, the shrinking berg, creaking and groaning, leaned precipitously to the right.

* * *

Bill Anders' iconic photo *Earthrise* shows our planet from 175,000 miles away. Seen from that distance, it looks like a fragile glass marble emerging from the nothingness of dark space. In 1968, the sight stunned the astronauts on Apollo 8 and the rest of the world. I was nineteen and I felt what I imagine many of us did: a momentary dropping away of conflict and strife, poverty, war, and injustice. I stared at the globe as if it were a tranquil jewel in the cold immensity of the galaxy. How, many of us wondered then, could we treat this singular, fragile entity with anything but grateful regard? Looking at that photograph, many of us had the first hint of what it might mean to feel a collective sense of responsibility.

Blue and alone for more than four billion years, without connection to other galaxies or stairways to heaven, without ideologies, cosmologies, or theologies of any kind, this was a place of intricate sustainabilities, which is a philosophical premise all its own and the starting point of an increasingly urgent kind of story.

From the cold dark of outer space, earth must also have felt to those astronauts like a silent magnet, a luring oasis, the one and only inhabitable place in the galaxy to which they could return and breathe, drink the water, live with other creatures.

* * *

What to do, then, when home is increasingly inhospitable? Fix it? Migrate? Haul out to another planet? That's the plan being not-so-secretly hatched by a group led by Elon Musk who've given up on earth and set their sights on starting over elsewhere. They're serious and rich. They held their inaugural "Mars Workshop" in August of 2018, though Musk has been working on this plan for years. Their goal: colonize Mars. Timetable: people on the red planet by 2024. They're steeped in stories like Noah's Ark.

* * *

The bergs were not companions that afternoon I kayaked in Tracy Arm Fjord, at least not in any human sense. Except for their creaks and groans as they shifted in the sea, they were silent and carried no angst about the future. Their end would come sooner than mine, but for the hour that day when I drifted among that pinnacled fleet with its histories of grandeur dwindling and soon to be gone, their diminishment seemed like mine, like ours. They had carried a certain beauty and power; they broke up, they broke apart. They lost whatever dignity they might have once had. They were about to vanish.

* * *

Glacial ice is roughed up and worn down by every story it encounters, including its own. Conventional wisdom says two thirds of bergs lie below water. The longer they float down the fjord toward the sea, the more they melt and the higher they rise. Their insides are complicated. Small millwells--slick-sided, circular drains--form to carry away the melting surface water through interior fractures.

As I watched that day in Alaska, another berg, big as a car, floated by under the mostly cloudy sky. For a moment, the sun's rays hit it just so and it sparkled. And then the sky darkened and the berg became just a berg again, shrinking and squeaking as it headed out to sea.

Lamentation

14

And then we will say the words and the words, hard-won and hefty, will become as if stones, and we will hold them before us and they will be our rock-solid refusal, our protest, our defiance and our sorrow. It won't matter if we are eighty or ten, whether we walk into rivers or streams or the sea itself; we will place those stones on the bottom and the stones will change the currents, alter the course, agitate the water enough to whip up an eddy that's never been there before, which is how all great turnings begin.

Cracks

If it hadn't been daylight at midnight, I might have seen the North Star almost directly overhead. At the end of an imaginary line that extends the earth's axis, the star, also known as Polaris or the Pole Star, is the brightest star in the Ursa Minor constellation, the one around which the night sky seems to swirl. In early maritime history, the North Star was a navigational guide. In western cultures today, it's a metaphor for a purposeful mission, a beacon of inspiration. In Norse mythology, it's pictured as the end of a spike. In Mongolian, it's a peg. Its function, many believed, was to hold the world together.

And now the polar icecap directly below it is coming apart.

Sometimes you can actually see the cracks begin—from the foredeck of an expedition ship, for example, off the northwest coast of Svalbard, Norway, where I'd recently gone to study the ice. One morning there I leaned over the railing and watched a skinny dark line in the ice appear at the bow of the boat. The tentative slit inched haltingly ahead; and then, more sure, it zig-zagged, widened, shoving away the white floes on either side. It spawned eight more cracks and a series of channels through which the sea began to surge. It was like watching a massive, top-of-the-world platform shatter in front of my eyes.

Through sea ice already thin enough in June for an ice-strengthened ship, we crunched our way north, aiming for 80° latitude, 650 miles from the North Pole. Scientists estimate that the ice pack there has thinned to less than half its 1980 thickness. It breaks more easily, melts more quickly. I take it as a given that most mythic destinations dissolve under scrutiny. But the pack ice here is more than myth. The pack ice is literal and its break-up beneath the North Star will do far worse than subvert anyone's notion of missions or guides.

* * *

The Arctic, most scientists agree, is warming twice as fast as any other place on earth.

They put part of the blame on the "albedo effect," the whiteness of a surface, which ranges from 0 (very black) to 1 (very white). An albedo of zero means the surface absorbs all incoming energy. The sun's rays get drawn in; the dark surface holds it. An albedo of one means the surface reflects all incoming energy. The sun's rays bounce back into the atmosphere. An albedo of zero means the surface has likely been warmed. An albedo of one means it's remained cool. Think of walking barefoot on black tar and then on snow.

The ice which once covered most of the Arctic region has an albedo of .6. It sends back 60% of incoming energy. Thick sea ice covered in snow returns 90%. The dark Arctic water has an albedo of .006. It reflects a mere 6%. Picture the loop: Big swaths of the once-massive cap reflected more than half the sun's energy, boomeranging it back up into the atmosphere. That reflection acts like a kind of air-conditioning, keeping the Polar Regions cool and therefore stable. But as more greenhouse gasses trap more warm air, more ice begins to melt. Less ice means less refraction. Less ice means more dark Arctic water exposed, which means more absorption of heat, which warms the waters which melts more ice. The loop becomes more than a loop. It becomes a runaway downward spiral. The pack ice has shrunk more than 300,000 square miles in recent years. The water, dark and absorbing, has increased that much.

It's not just as if that the air-conditioning has been turned off, but that the heaters are now on.

* * *

Sometimes the cracks came quickly, and a solid swath of white smithereened into chunks that floated away on the cold, dark sea.

Sometimes they happened slowly—a single skinny line etched lightly, the floe left almost intact.

* * *

Sometimes you can watch a different kind of crack begin--in allegiance, say, to a story you once loved. Jesus fed the multitudes and raised Lazarus from the dead. A miracle meant the laws of nature had been superseded, that the facts of too few fish or loaves of bread or the death of brain cells meant nothing compared to cherished beliefs. But then comes news of famines and the ravages of disease, plagues and wars started by madmen, and the slow realization that those "laws" of nature are not laws after all, not decreed by anyone or handed down or voted on. They are, rather, the unavoidable, inexorable patterns of cause and effect re-asserting themselves. And so those miracles come under closer scrutiny, where they can't hold up. I remember the day that I, a child besot by all things biblical, stepped off the bank of a pond and swore to God if he'd keep my feet dry I'd devote my whole life to him. But physics is physics. How else to understand the fact of soaked shoes, the sobering re-calibration of trust? Looking back, I can see in that day by the pond the initial rifts, the early fault lines etched in what had been my bedrock foundation.

* * *

Just before noon, we crossed the 80° latitude line. Among the crew and passengers, there was much cheering and high-fiving, as if we had accomplished some elusive goal, and indeed, the ship had never been that far north. The ice would not have allowed it.

We know from oral histories and adventurers' logs that the Arctic has, for eons, been covered with ice year round. Most of that was "old ice" that thickened year after year. Some of it was "new ice" that formed each winter and melted each summer. Today the ratio of old ice to new has drastically changed. Though 30% of the ice cap is still old and thick, that percentage is dropping ever year. The remaining 70%, a figure that's increasing every year, is new and, almost by definition, thinner, more subject to warming currents. I don't know how to feel triumphant about crossing that latitude line.

* * *

At the entrance to Raudfjorden, the ship, surrounded now by thicker ice, slowed to a halt. The guides had agreed to let us hike out onto the floe. They lowered the zodiacs and issued the warnings: no stepping beyond a red line they'd marked in the ice and now guarded with guns. If you see a polar bear, don't run.

I couldn't help but think of those moments two years ago in Hawaii when my granddaughters and I held hands and somersaulted underwater and I could still imagine the coral reefs in recovery. How different the unforgiving north is where you can't swim, can't, in fact, go anywhere without cold-weather gear, life jacket, ear muffs, water-proof boots so heavy you have to haul your legs awkwardly over the sides of the raft. You trudge out on the ice stiffly, armored against the weather. Your skin seems faraway.

So does the rest of the world. No roads or cell phones or Internet. When fog shrouded the pack ice, cloudy skies faded into silvery ice, barely a line between them. Visibility shrank. No North Star for positioning, no landmarks, buoys, or foghorns, nothing that might help you situate your body, tell you where you are.

Sometimes I wonder what kind of unconscious trajectory I've been on for the last two years, starting in the tropics and ending up here. If there's a progression to feeling the effects of the climate crisis, maybe it's been propelling me northward all along, depositing me here at last in this austere and exacting land at the top of the world so that I could feel in my bones what it's like when alternatives disappear yet the way ahead remains so unclear. Even the ship, a haven I knew was anchored out in the fjord, looked like a ghost ship, appearing and disappearing in the fog.

Dear god, I might have once begun.

* * *

There are no indigenous people in Svalbard, no long traditions rooted here. In other parts of the High Arctic, the Aleut and Saami, the Nenets and Inuit have, for ages, been telling stories about sustaining relationships with the land and the web of ancestral families. But though the Vikings passed through Svalbard, along with the Russians, the Dutch, Norwegians, and others, there have been no ancestral homes here and thus no time for story-telling over generations, no accumulating of tales.

Here in the bleakness of wind and snow and frozen ground, there have been only the polar bear and reindeer, arctic fox, and the bright orange beak of the puffin—each with its survival strategies and small dependencies that creatures work out over millennia. But no histories, oral or written, have embedded the realities of their lives—their umwelt—in the human imagination, and so nothing here can extend our empathy or undo the futility of trying to pass down what remains unembedded and rootless.

To be without a story, Rebecca Solnit claims, *is to be lost in the vastness of a world that speeds in all directions like arctic tundra or sea ice.* To be in the High Arctic, is not, it seems to me, to be without a story but to feel stranded in the diminishing power of stories that used to matter, to see that a story that implies a promise is somehow a lie, an illusion that will show the story itself will ultimately be betrayed. Those beloved miracles I once clung to have little to do with the polar bear we'd seen this morning or the seal who slipped off an ice floe just in time. Or with migrations spurred by drought or whole species' extinctions. Or, more pointedly, with tragedies triggered by our own greed, the degradation that will hit the most vulnerable people first.

What morality tale matters now? What moral compass? What star to steer by?

When traditional stories no longer apply, some of us sometimes turn to more modern masterpieces that identify our existential crises and offer, if nothing else, the balm of companionship. They amuse us, for example, with the absurdity of living only to wait for salvation from outside human activity. They remind us that we'll stay on the stage because that's what we do, because we're willing to live as if sooner or later Godot will actually come along.

We tell those stories, we see those plays, we walk away comforted by dark visions like Beckett's, maybe even exhilarated by having felt seen by an artist or just by thinking that, at least in art, there's always another show.

But up here in the Arctic, even those stories don't satisfy. It's as if the landscape insists they be starker. If we staged Beckett's play in some Arctic theater, it would have to have a different ending: Maybe Kafka enters and announces to Vladimir and Estragon and to all of us his famous line, "Oh plenty of hope, an infinite amount of hope—but not for us." Then he and the two waiting men climb down off the stage and join the rest of us in our seats. The theater's doors disappear; there's nothing to do but sit there with one another in the fog.

Maybe that would be the truer play.

* * *

Back aboard the ship a few hours later, I leaned over the rail again and watched as those dark, sinuous lines multiplied right in front of me, spider-webbing, as if the ice were nothing more than a floating mosaic, flimsy pieces jostled by warming and pulling apart so that you can see the dark water, clear as a channel, which then closes. How does the captain navigate through a broken-up ice pack? How does he know which opening will stay open, let the ship pass through? How do any of us know which path forward will close in soon, leave us stranded in a dead end?

Up on the bridge, I watched him scan the instruments that tell him exactly where we are, how deep the water, how extensive the ice. He checked and re-checked multiple screens, including one that gives him a digital picture of the ice ahead, where it's thin enough to break through, what kind of open water lies where and how to steer us there.

I envied him his instruments and his certainties.

* * *

In Martha Graham's masterpiece titled "Lamentation," a single figure wrapped in a stretchy cloth squirms for four minutes on a bench, leaning left then right, reaching up, bending over, as if she cannot abide what's twisting inside her, as if she's trying futilely to writhe into relief. The dance is not meant to convey particular mourning for particular loss but to embody archetypal grief itself.

Here's how it feels, Graham seems to say: You do not walk away or deny or give up. You're stuck on a bench, wrapped in a purple shroud that moves with every move you try to make, transforming every distorted thrash into a visible line, a traceable shape.

* * *

It's not just that the ice is disappearing, one of the guides reminds me. Thawing permafrost is even worse. It's a ticking bomb. Schymon sank into a chair without his gun and Arctic garb. The pack ice had opened up again; the ship was underway. I'd been in southern Svalbard, in Longyearbyen, the week before. You can see there the tilted gravestones and crumbling pavement caused by melting permafrost. Some of the roads undulate like tattered ribbons. You can feel the unsettling under your feet and in a local scientist's prediction about the slumping hillside above town where an

146

avalanche recently barreled into a cluster of homes below, killing a man and a child. "It'll happen again," he declares. "We need an early warning system."

But in places like Siberia, Schymon says, shaking his head, the permafrost is much thicker than it is here, much older, much denser with organic rot, and it's melting much faster.

The result is yet another feedback loop on the verge of a tipping point. The ancient biomass—trees, grasses, animal remains, anything that lived in the Arctic-- collapsed, as all living things eventually do, and was eaten and discarded, subjected to the usual cycles of microbial decay. But anything half-decomposed when a period of prolonged cold began remained suspended in that state of partial decay and was buried under succeeding depositions, which the Ice Ages chilled again and again. Into the dark matter the carbon-loaded stuff sank deeper and deeper, was compressed under pressure, and finally frozen solid. And there it lay, inert, inedible, unappetizing for thousands of years. Scientists estimate that the amount of carbon stored in permafrost worldwide is 1,400 gigatons—way more than is already in the air.

In a climate of global warming, that frozen ground has begun to thaw. Where that permafrost is chock full of organic dregs and heated now enough for devouring, the microbes settle in to do what microbes do: feed in a frenzy that releases carbon dioxide and methane. More of those gasses released into the atmosphere means more warming, which thaws more permafrost, which means more decay and CO2 and methane. Yet another unstoppable cycle begins. In Siberia, Schymon says, methane, which can trap twenty times more heat in the atmosphere than carbon does, is starting to bubble up through lakes.

* * *

When the curtain dropped at the end of one of Martha Graham's performances, a woman in the audience approached the artist. Her child, she told Graham, had recently been killed in an accident that she, the mother, had witnessed. She had not been able to mourn. Seeing *Lamentation* opened up what had been locked down. The woman wept. It wasn't, evidently, just that the dance was so visceral. It was that her sorrow had become a "dignified and valid emotion." That recognition, I still believe, is one of the powers of story and art.

* * *

I don't know how a polar bear survives if it's stranded on a shrinking bit of ice.

I don't know whether every shattered bit of an outdated story still carries in it something that was once alive with a truth that might yet be resurrected in a newer kind of story.

I don't know how to think about the future.

I do know that the old Western stories are failing us now. For centuries, they have fed our psyches with a questing spirit and notions of triumph and resolution, with endings that satisfy because they give us hope, which inspires action, the can-do spirit in full can-do mode, which is another way of saying they give us a way to avoid our griefs. Their failure to help us navigate the wildness of what's happening now will be yet another failure of our imaginations, akin to the failure that got us into this mess in the first place.

* * *

The end will not likely begin with an earthquake. That sudden drama and all the protocols of rescue and recovery we've honed over the past hundred years will be irrelevant. The Arctic, once the mythic land below the guiding North Star, is marred by fractured ice and seeping gasses. We ignored its early warning signs for decades and now the direction is clear: where we're headed is straight into a massive splintering of who and what and where we think we are.

This, I realize, is how it feels to drift in the epilogue of a big and once-beloved story whose ending has been shattered. We are stragglers, now, all of us, and the urgent questions before us seem less like ones about reversals and overcoming, or the absurdity of perpetual waiting, and more like the final line from a Robert Frost poem:

what to make of a diminished thing.

* * *

If it's too late for the well-loved stories of heroism and triumph, for ingenious structures and satisfying resolutions, or for irony, or the tactics of unreliable narrators, what then should our stories be? Or maybe the better question is this: what human impulses do we want to guide us on this final journey?

Generosity. Compassion. A willingness to immerse ourselves here, to be with all of our dying. If there's anything redemptive about what's likely to happen, perhaps it's in our learning to grieve in the gritty chaos together. To fashion new stories not of action but of intermingled attentions. They will embody the hard work of noticing, of not turning from but abiding with. Hospice stories. When a friend of mine lay dying, she wanted nothing more but for those of us who loved her to be nearby. We sat by her bed; we held her hand. What we said or didn't say mattered less than the manners with which we attended her dying. As she made her way into absence, what mattered was our presence.

Off the northwest coast of Svalbard, I watched another chunk of ice from a massive floe drift southward. The water had softened its edges, turned them to mush, and lapped into the melt-pond in middle. The sun's rays no longer bounced back. Eventually the chunk will shrink to almost nothing, and then the sea will absorb what remains. It will be as if the ice had never been there.

Sixth Extinction | Holocene Epoch/Anthropocene
11,700 years ago to Present

Once the dinosaurs were gone, small mammals had a chance. Some of them became big animals, some moved out of forests, crossed continents. Some, like wolf-like mesonychids, abandoned dry land and evolved into whales. Human-like primates emerged. Some grew larger, began to walk upright, shed their fur and grew bigger craniums for bigger brains. Some became us.

For hundreds of thousands of years, this was a time of little drama, geologically speaking. The large land masses settled into something close to their current positions. There were no massive extinctions, no sudden catastrophic disruptions that killed off entire species overnight. A couple of million years ago, the planet's tilting towards then away from the sun set off a series of Ice Ages. Large mammals and the plant life they browsed on migrated north then south as the ice retreated and advanced.

The last Ice Age ended some 11,000 years ago. Since then we humans have multiplied and spread across the planet, developed language, art, religion, and the capacity for compassion, imagination and ingenuity.

In approximately the same time span, 90% of the huge mammals in North America disappeared. Some speculate about climate change. Others blame hunting by humans who, furless, also figured out how to unearth the millions of years of fossil fuels buried deep in the earth and burn them for warmth, for industry and transportation.

We burned and burned and burned and CO2 levels in the atmosphere rose. And then came half a billion more people every decade and with every decade more deforestation, more habitat destruction, more burning of fossil fuels.

CO^2 levels, already higher than they'd been for at least 800,000 years, rose a hundred times faster than they ever had, and that meant seas warmed; coastal lands flooded. Wildfires scorched and drought parched. Massive human migrations began.

In the early 21st century, species extinction rates soared 100 to 1000 times higher than the normal background rate, rising exponentially faster than ever before. A million species stood on the brink of disappearing forever.

And then the permafrost which had locked up carbon and methane for thousands of years began to melt.

And so it was that another perfect storm approached.

Lamentation

15

From here on out, any stories that endure must begin with an ancient hiss from sea-bottom vents, swelling of seeds and small lungs, the blood ark burrowing into sand, anything willing to begin again, to live fiercely in cahoots with the minutiae of give and take

and the dances that matter most must be those akin to the manta ray's mouth-open somersault, the dolphin's leaping spin, the whale breach, and slow, sand-hugging slide of the moon snail, any move, including death, that means creaturely return to the elemental.

The gods to honor now are the ones who emerge from sea bottoms and tree tops, embed themselves like clownfish among poisonous anemones, bare their teeth as the urchins do on bony coral, promise nothing but more of the earthy links and collusions that keep the whole shebang banging.

The rites to preserve from here on out are the ones that subvert any attempts to steal from the future, and the liturgies to exalt are those that evolve from web within web, nest within nest, as it was in the beginning, is now, and forever shall be,

and the congregations to embrace are the ones that keep baptizing us in the name of the mountains, the soil, and steamy rock, wild oceans and stars overhead,

all in the name of the earth itself, the spinning, ever-sprouting cornucopic globe that once upon a time was a world uncorrupted by an excess of gluttony and greed

and that remains our only home where all that might be left for us is to walk these disappearing places, to be in and among, in the thick of, ensnared by and up to our necks in everything here, in everything we are and in everything we're about to destroy

and then the world will become, as it has always been becoming, a different world.

**Barbara Hurd** is the author of *Listening to the Savage / River Notes and Half-Heard Melodies (*University of Georgia Press 2016), *Tidal Rhythms* (with photographer Stephen Strom, George F. Thompson Publishing 2016), *Walking the Wrack Line* (University of Georgia Press 2008), *Entering the Stone*, a Library Journal Best Natural History Book of the Year (Houghton Mifflin 2003), and *Stirring the Mud,* a *Los Angeles Times* Best Book of 2001 (Beacon Press 2001). Her work has appeared in numerous journals including *LitHub, Bellingham Review, Prairie Schooner, Best American Essays, The Yale Review, The Georgia Review, Orion, Audubon, The Sun*, and others. The recipient of a 2015 Guggenheim Fellowship, an NEA Fellowship for Creative Nonfiction, winner of the Sierra Club's National Nature Writing Award, three Pushcart Prizes, and five Maryland State Arts Council Awards, she teaches in the MFA in Writing Program at the Vermont College of Fine Arts. *www.barbarahurd.com*